AF428694

PARALLEL LIVES, PARALLEL REALITIES

Parallel Lives, Parallel Realities

Unveiling the Multiverse

DEMETRI WELSH

DemetriWelsh.com

CONTENTS

~ 32 ~

Conclusion: Integrating the Multiverse into Our Lives

~ 33 ~

Appendix: Further Reading and Resources

~ 34 ~

Glossary of Terms

To my dearest husband Anthony,

In you, I have found an unwavering pillar of support, a fellow scribe in the boundless realms of creativity, and a mirror to the best parts of myself. Your brilliance as a public author not only illuminates your own path but also lights the way for me and countless others fortunate enough to traverse alongside you.

Your strength, wisdom, and compassion remind me daily of the greatness to which we all can aspire. This book, while a journey into the many worlds beyond our reach, is grounded in the reality of my admiration and respect for you—my partner, my muse, my guide.

With all my love and deepest gratitude,
Demetri Welsh

PROLOGUE

Welcome, dear reader, to a journey that stretches beyond the boundaries of our known universe, into realms that challenge the very fabric of reality. I am Demetri Welsh, and I will be your guide through the pages of this exploration, a narrative that delves into the existence of parallel lives and dimensions. Each page, each chapter, is a door swinging wide into possibilities both strange and wondrous, a gallery of worlds that might exist alongside our own.

Our quest begins at the frontier of quantum mechanics and string theory, where the seeds of the multiverse theory take root in the fertile soil of modern physics. These scientific theories suggest a cosmos brimming with alternate versions of reality—some nearly identical to our own, others drastically different in ways that defy imagination. This book is an homage to the potential infinity of universes, an endeavor to explore the ramifications of such an expanse on our understanding of existence.

But why venture into these speculative realms? The answer lies in the innate human curiosity, the drive to look beyond the horizon and ask "What if?" What if somewhere, in some other layer of reality, different versions of us lead lives unrecognizable to our own? What if, by exploring these possibilities, we could learn something profound about the nature of reality itself and our place within it?

As a metaphysician and psychic reader, I have always been drawn to the mysteries that lie at the edges of science and spirituality.

This book merges those worlds, weaving together the threads of theoretical physics and metaphysical speculation to craft a tapestry of multidimensional exploration. Each chapter is a reflection not only of a universe that might be but also of the human condition—our fears, our hopes, and our endless quest for knowledge.

Throughout these pages, I invite you to suspend disbelief and embrace the wonder of the unknown. The dimensions we will explore are constructed from a blend of researched facts, educated guesses, and imaginative leaps. From worlds where time flows backward to realms where light is the substance of life, each fictional universe is a case study in what might lie beyond the veil of our current understanding.

In dedicating this book to my beloved husband, a fellow author and an unwavering source of inspiration, I am reminded of the power of partnership in the face of the unknowable. His belief in the potential of ideas and the importance of sharing them has propelled this endeavor from mere speculation to a vivid narrative that I now share with you. It is my hope that within these explorations, you find not only entertainment but also food for thought—concepts that challenge you, inspire you, and perhaps change how you view the cosmos.

Let us embark on this adventure together, with open minds and eager hearts, ready to discover the myriad dimensions that might parallel our own. As we turn each page, let us wonder, let us question, and let us dream of the infinite universes that could exist beyond the stars.

Welcome to *Parallel Lives, Parallel Realities: Unveiling the Multiverse.* The journey begins now.

Unveiling the Foundations

Welcome to the first segment of our journey through the vast and varied landscapes of the multiverse. I am Demetri Welsh, and in Part One of *Parallel Lives, Parallel Realities: Unveiling the Multiverse*, we will lay the groundwork for our exploration into the concept of parallel universes—a foundational dive into the theories and ideas that make such extraordinary speculations possible.

The Quest for Understanding

Human curiosity knows no bounds. From the earliest days of civilization, we have looked up at the stars and wondered about the nature of the cosmos. What lies beyond the visible? How vast is reality? Are we alone in the universe? These questions have propelled humanity's greatest scientific pursuits. Today, these inquiries push us toward the edges of theoretical physics and metaphysics, where the possibility of multiple universes suggests answers as profound as the questions themselves.

The Scientific Pillars

In this first part, we will explore two major scientific frameworks that have led to the multiverse theory: quantum mechanics and string theory. Each chapter will focus on a key aspect of these theories, breaking down complex concepts into understandable segments without sacrificing the depth of scientific rigor. We aim to make these theories accessible to all, regardless of scientific background, providing a clear understanding of how contemporary science allows for the possibility of countless other universes.

Quantum Mechanics: The Doorway to Possibilities

We begin with quantum mechanics, the branch of physics that deals with the smallest particles of the universe. Here, particles exist in states of potential, embodying multiple possibilities at once. This principle, when extrapolated, suggests that every quantum event could result in the branching of reality into multiple outcomes—each a separate, parallel universe.

String Theory: A Symphony of Dimensions

Following our exploration of quantum mechanics, we delve into string theory, which posits that the most fundamental components of the universe are not points, but strings that vibrate at various frequencies. String theory is crucial because it introduces the concept of higher dimensions—beyond our familiar three-dimensional space and one-dimensional time—and provides a framework for a universe much richer and more complex than we could observe with our senses alone.

Philosophical and Ethical Implications

Understanding the scientific underpinnings of the multiverse also demands that we confront its philosophical and ethical implications. What does it mean for our understanding of free will, identity, or fate if there are infinite versions of ourselves making different choices in parallel universes? How do we grapple with the moral ramifications of actions that spawn countless outcomes across multiple realities?

Setting the Stage

Each chapter in this first part is designed to not only inform but also to challenge you to think deeply about the nature of reality. As we dissect these theories, we'll uncover the building blocks needed to appreciate the vast potential of the multiverse, setting the stage for the more detailed exploration of individual alternate dimensions in the later parts of the book.

Together, let's embark on this intellectual adventure, equipped with the necessary tools to explore the furthest reaches of possibility. By the end of Part One, you will have a solid foundation in the theories that make the concept of parallel universes not just a science fiction trope but a frontier of modern scientific inquiry.

~ 1 ~

THE MULTIVERSE THEORY
EXPLAINED

Welcome, intrepid explorer of the unknown. I am Demetri Welsh, and in this opening chapter, we embark together on an exploration of one of the most profound and potentially transformative concepts in modern physics and metaphysics: the multiverse theory. Here, we will unravel the scientific underpinnings, philosophical implications, and the sheer scale of what it means to entertain the existence of multiple universes.

The Foundations of Multiverse Theory

At its core, the multiverse theory posits that our universe, with all its vast complexity and myriad forms of matter and energy, is not the only universe but just one of an enormous number, perhaps an infinite number, of universes that exist simultaneously. This is not merely a fantastical notion but one that stems from real, albeit mind-bending, developments in physics.

The seeds of this theory lie in quantum mechanics, the branch of physics that deals with the very small. Quantum mechanics has shown us a world where particles exist in a state of potential—multiple states at once—until observed. It suggests a fundamental randomness in the nature of reality, where probabilities replace certainties and observation itself shapes the outcome.

Quantum Mechanics and the Path to Parallel Universes

Consider the famous thought experiment proposed by physicist Erwin Schrödinger, wherein a cat in a sealed box can be simultaneously alive and dead, its fate tied to a random quantum event. This illustrates the principle of superposition, where systems are in multiple states at the same time. But where does each possibility go? The theory suggests that for every possible outcome that could come from one of these quantum processes, there might be a separate, parallel universe.

From quantum mechanics, the step to a multiverse isn't as large as it seems. The many-worlds interpretation, one of the most popular multiverse hypotheses, asserts that all possible alternative histories and futures are real and exist in parallel universes. Each time a quantum event has more than one possible outcome, the universe splits, creating a set of parallel universes, one for each possible outcome.

String Theory and the Landscape of Universes

Another scientific framework that lends credence to the idea of a multiverse is string theory. String theory proposes that the fundamental particles we observe are not point-like dots, but rather tiny, vibrating strings. These strings exist in a landscape of 10 or more dimensions, far beyond the four-dimensional spacetime we experience daily. According to string theory, different ways these strings can vibrate correspond to different fundamental particles. The theory also allows for a vast number of possible configurations of these dimensions, each configuration potentially representing a different universe with its own unique set of physical laws.

Philosophical Implications of the Multiverse

The philosophical implications of the multiverse theory are as profound as the scientific assertions. If multiple universes exist, what does that say about our notions of reality, fate, and free will? Are there versions of us making different choices, living different lives? What might it mean for moral and ethical decisions if every choice spawns a universe where the opposite choice is made?

Bringing It All Together

As we contemplate these questions, it becomes clear that the multiverse theory isn't just about physics; it's about the very nature of reality, identity, and the human condition. It challenges us to expand our thinking beyond the universe we see, to consider realities that might be radically different from our own.

In this book, we'll explore these realities, diving deep into the realms of what might be, powered by the twin engines of scientific inquiry and boundless imagination. As we move forward, keep in mind that these ideas, while speculative, are rooted in the latest developments and theories in modern science—they invite us to look at the universe in entirely new ways.

So, as we turn the page to the next chapter, let us carry with us an openness to the vast possibilities that lie ahead, emboldened by the knowledge that in the quest to understand our universe, the only limit is the extent of our curiosity. Let's journey together through the cosmic landscape of the multiverse.

~ 2 ~

QUANTUM BEGINNINGS

In this chapter, we delve into the intricate world of quantum mechanics, a pillar of modern physics that underpins much of our understanding of the multiverse theory. Quantum mechanics is not just a theory about how things work at the smallest scales of energy and matter; it is a revelation about the fundamental nature of reality itself. Let us explore these quantum beginnings and how they lead us to consider the existence of parallel universes.

The Birth of Quantum Mechanics

Quantum mechanics began to take shape in the early 20th century, when scientists started to explore the atomic and subatomic levels of matter. They discovered that the laws of classical mechanics, which worked so well in describing the motion of planets and everyday objects, failed when applied to the atomic scale. This realization led to a revolutionary new framework, fundamentally altering our understanding of nature's rules.

Key Experiments and Findings

Planck's Quantum Theory: The journey into quantum mechanics began with Max Planck's solution to the black-body radiation problem in 1900. Planck proposed that energy is emitted and absorbed in discrete quantities, which he called "quanta." This was a radical departure from classical notions of energy as a continuous flow.

- Planck's Quantum Theory marks a pivotal moment in the history of physics, heralding the birth of quantum mechanics and challenging the very foundation of classical physics. Max Planck's groundbreaking work emerged from his attempt to resolve a perplexing anomaly in classical physics: the black-body radiation problem.

Classically, according to the prevailing wave theory of light, it was assumed that energy could be emitted or absorbed continuously, without any restrictions. However, when scientists attempted to model the emission of electromagnetic radiation from a black body (an idealized object that absorbs and emits all frequencies of electromagnetic radiation), they encountered a significant discrepancy between theory and observation. Classical physics predicted an infinite amount of energy at short wavelengths, an absurdity known as the ultraviolet catastrophe.

In 1900, Planck proposed a radical departure from classical physics by suggesting that electromagnetic energy is not emitted or absorbed continuously, but rather in discrete packets or "quanta." These quanta, or bundles of energy, were a revolutionary concept, challenging the prevailing understanding of the nature of light and energy. Planck introduced a fundamental constant, now known as Planck's constant (h), to quantify these discrete energy packets.

The implications of Planck's quantum hypothesis were profound. It not only provided a solution to the black-body radiation problem, but it also laid the groundwork for a new branch of physics: quantum mechanics. The notion of energy quantization introduced by Planck paved the way for further revolutionary discoveries by Albert Einstein, Niels Bohr, Werner Heisenberg, Erwin Schrödinger, and others.

Planck's Quantum Theory fundamentally altered our understanding of the behavior of matter and energy at the microscopic level. It shattered the classical notion of determinism and introduced the probabilistic nature of quantum phenomena. Planck's work marked a paradigm shift, transforming physics and leading to the development of numerous technologies, including semiconductors, lasers, and quantum computing, that have revolutionized the modern world.

The Photoelectric Effect: Albert Einstein furthered Planck's ideas by explaining the photoelectric effect, for which he won the Nobel Prize in Physics. Einstein proposed that light itself is made up of packets of energy, or photons, which exhibit both wave-like and particle-like properties.

- The photoelectric effect, one of the most significant experimental phenomena in the realm of quantum physics, played a pivotal role in reshaping our understanding of light and matter. Albert Einstein's groundbreaking explanation of this phenomenon not only consolidated Max Planck's earlier insights but also laid the foundation for the concept of photons and the dual nature of light.

  First observed by Heinrich Hertz in 1887, the photoelectric effect refers to the emission of electrons from a material's surface when it is exposed to light. Classical wave theory struggled to explain certain aspects of this effect, such as the dependence of the emitted electron's kinetic energy on the frequency of incident light rather than its intensity, as well as the instantaneous emission of electrons upon exposure to light.

In 1905, Albert Einstein provided a revolutionary explanation for the photoelectric effect in one of his seminal papers, postulating that light behaves not only as a wave but also as discrete packets of energy, which he termed "quanta" or "photons." According to Einstein's theory, when light interacts with matter, such as a metal surface in the case of the photoelectric effect, it transfers its energy to individual electrons in the form of photons. If the energy of these photons exceeds the material's binding energy for electrons, the electrons can be ejected from the surface.

Einstein's explanation of the photoelectric effect reconciled several experimental observations that classical wave theory couldn't account for. Specifically, it explained the observed threshold frequency phenomenon, where electrons are only emitted above a certain frequency threshold, as well as the direct proportionality between the kinetic energy of emitted electrons and the frequency of incident light. Moreover, Einstein's theory predicted that increasing the intensity of light would only increase the number of emitted electrons, not their kinetic energy—a prediction that aligned perfectly with experimental observations.

Einstein's groundbreaking insights into the photoelectric effect not only validated the concept of photons but also provided compelling evidence for the wave-particle duality of light. This duality suggests that light exhibits both wave-like and particle-like properties, depending on the context of observation. Einstein's work on the photoelectric effect earned him the Nobel Prize in Physics in 1921 and further cemented the foundations of quantum theory, revolutionizing our understanding of the fundamental nature of light and matter.

Heisenberg's Uncertainty Principle: Werner Heisenberg introduced the Uncertainty Principle, which states that the more precisely the position of some particle is determined, the less precisely its momentum can be known, and vice versa. This principle challenged the deterministic nature of classical physics, introducing a probabilistic framework to the atomic world.

- Heisenberg's Uncertainty Principle stands as a cornerstone of modern physics, fundamentally altering our perception of the microscopic world and challenging the deterministic worldview of classical physics. Introduced by Werner Heisenberg in 1927, this principle revolutionized our understanding of the limitations of measurement and the inherent uncertainty present in quantum mechanics.

  At its core, the Uncertainty Principle asserts that there are fundamental limits to the precision with which certain pairs of physical properties, such as the position and momentum of a particle, can be simultaneously measured. Specifically, it states that the more precisely we know the position of a particle, the less precisely we can know its momentum, and vice versa. This principle arises from the wave-particle duality inherent in quantum mechanics, where particles exhibit both wave-like and particle-like behaviors.

  Heisenberg's profound insight stemmed from the realization that in order to observe a particle's position, one must interact with it in some way, typically through the use of electromagnetic radiation like light. However, this interaction inevitably disturbs the particle's momentum, making it impossible to precisely determine both properties simultaneously. This inherent uncertainty is not due to limitations in measurement techniques but is instead a fundamental aspect of the quantum nature of reality.

The implications of the Uncertainty Principle were profound and far-reaching. It shattered the classical notion of determinism, which posited that the future behavior of a system could be precisely predicted given complete knowledge of its initial conditions and governing laws. Instead, Heisenberg's principle introduced a probabilistic framework to the atomic world, where the behavior of particles is inherently uncertain and governed by probabilities.

Heisenberg's Uncertainty Principle played a pivotal role in shaping the development of quantum mechanics and our understanding of the microscopic realm. It paved the way for further groundbreaking discoveries, such as the development of quantum field theory and the concept of entanglement. Moreover, it led to the recognition that the act of observation itself can profoundly influence the behavior of quantum systems, giving rise to the philosophical debates surrounding the nature of reality and the role of the observer.

In summary, Heisenberg's Uncertainty Principle represents a fundamental shift in our understanding of the universe, challenging classical notions of determinism and introducing a probabilistic framework to describe the behavior of particles at the quantum level. It remains one of the most profound and enduring principles in modern physics, shaping our understanding of the fundamental nature of reality.

Schrödinger's Wave Function: Erwin Schrödinger developed the wave function, a mathematical description of the quantum state of a system. The wave function evolves deterministically but describes probabilities of outcomes, not certainties.

• Erwin Schrödinger's development of the wave function stands as a monumental achievement in the field of quantum mechanics, providing a powerful mathematical framework for describing the behavior of particles at the microscopic level. Schrödinger's wave function revolutionized our understanding of quantum systems by offering a deterministic evolution of the system's state while simultaneously acknowledging the inherent probabilistic nature of quantum phenomena.

At its essence, the wave function is a mathematical function that describes the quantum state of a system. For a single particle, the wave function $\Psi(x, t)$ specifies the probability amplitude of finding the particle at a particular position x at a given time t. Schrödinger's famous equation, the Schrödinger equation, governs the evolution of this wave function over time. The equation represents a fundamental principle in quantum mechanics, describing how the wave function changes in response to the system's energy and potential.

One of the key insights of Schrödinger's wave function is its deterministic evolution. Given the initial conditions of a quantum system and the governing laws described by the Schrödinger equation, the evolution of the wave function is entirely deterministic. This deterministic nature allows physicists to predict the future behavior of quantum systems with remarkable accuracy.

However, despite its deterministic evolution, the wave function does not provide certainties about the outcomes of measurements. Instead, it encodes probabilities of different outcomes. According to the interpretation of quantum mechanics pioneered by Niels Bohr and others, the square of the magnitude of the wave function, $|\Psi(x, t)|^2$, gives the probability density of finding a particle at a particular position x at time t. This probabilistic interpretation of the wave function reflects the inherent uncertainty present in quantum systems, as described by Heisenberg's Uncertainty Principle.

The probabilistic nature of the wave function has profound implications for our understanding of the microscopic world. It implies that in the realm of quantum mechanics, we can only make statistical predictions about the outcomes of measurements, rather than precise determinations. This departure from classical determinism underscores the fundamentally probabilistic nature of quantum phenomena and highlights the limitations of our ability to precisely know the properties of particles at the quantum level.

Despite its probabilistic interpretation, Schrödinger's wave function has proven to be an indispensable tool for physicists in understanding and predicting the behavior of quantum systems. Its deterministic evolution combined with its probabilistic interpretation has enabled remarkable advances in fields such as quantum chemistry, quantum computing, and quantum information theory. Schrödinger's wave function remains a cornerstone of modern quantum mechanics, providing a powerful mathematical framework for describing the complex and fascinating behavior of particles at the smallest scales of the universe.

Quantum Superposition and Entanglement

One of the most startling findings in quantum mechanics is the principle of superposition, where particles can exist in multiple states simultaneously until measured. The famous thought experiment of Schrödinger's Cat illustrates this: a cat in a box with a quantum mechanism can be considered both alive and dead until the box is opened and an observation made.

Entanglement, another quantum phenomenon, occurs when pairs or groups of particles interact in ways such that the quantum state of each particle cannot be described independently of the state of the others, even when the particles are separated by large distances. This suggests a fundamental interconnectedness of quantum particles, regardless of the space between them.

From Quantum to Cosmos

The implications of quantum mechanics extend beyond tiny particles. They suggest that at every moment, countless interactions and decisions at the quantum level could lead to the branching of universes—a key idea in the multiverse theory. Each possible outcome of a quantum event could correspond to a different "world" in a vast multiverse.

The Many-Worlds Interpretation

Hugh Everett III proposed the Many-Worlds Interpretation (MWI) of quantum mechanics in the 1950s. According to MWI, all possible alternate histories and futures are real and exist in a vast multiverse of universes. This interpretation removes the need for wave function collapse, which many find problematic, and instead postulates that all possibilities inherent in quantum states are physically realized in some set of universes.

Conclusion

Quantum mechanics has fundamentally changed our understanding of the natural world, suggesting that reality is not as fixed or as straightforward as it appears. Its principles challenge our classical intuitions about how the world works and open the door to a universe—or rather, a multiverse—where the incredible becomes possible. As we contemplate these quantum beginnings, we lay the groundwork for exploring deeper into the nature of existence and the many realms that may parallel our own.

~ 3 ~

HISTORICAL PERSPECTIVES ON PARALLEL REALITIES

In this chapter, we explore the fascinating journey of human thought regarding the concept of parallel universes and multiple realities through various historical and cultural lenses. This exploration is not only about understanding modern physics but also about appreciating how ancient philosophies and medieval beliefs contribute to our contemporary discussions about the multiverse. Let's delve into how cultures across time have envisioned other worlds and realities.

Ancient and Pre-Modern Concepts of the Cosmos

Multiple Worlds in Mythology: Many ancient cultures had myths that spoke of other worlds and realms. For example, the Norse cosmology includes nine worlds, each home to different beings like gods, giants, and humans, interconnected by the world tree, Yggdrasil.

- The concept of multiple worlds, alternate realms, or parallel universes is a recurring theme in mythology and folklore across various cultures throughout history. These myths often reflect humanity's fascination with the unknown and the desire to explore realms beyond our own. One prominent example of such mythology is found in Norse cosmology, which encompasses a rich tapestry of interconnected worlds inhabited by diverse beings.

In Norse mythology, the cosmos is depicted as a complex network of nine worlds, each with its own distinct characteristics, inhabitants, and significance. These worlds are interconnected by the cosmic ash tree, Yggdrasil, which serves as the axis mundi, or the central pillar of the universe. Yggdrasil's branches extend across the cosmos, linking the various realms together and facilitating communication and travel between them.

At the heart of Norse cosmology lies Asgard, the realm of the Aesir gods, who govern aspects of the natural world, warfare, and fate. Asgard is depicted as a celestial stronghold, home to majestic palaces and halls where the gods reside. Surrounding Asgard are several other realms, each with its own inhabitants and mythology:

1. Vanaheim: Home to the Vanir gods, associated with fertility, agriculture, and nature.
2. Alfheim: Inhabited by the light elves, beings of beauty and grace who dwell in radiant realms.
3. Midgard: The realm of humanity, situated at the center of the cosmos and surrounded by an impassable ocean.
4. Jotunheim: The land of the giants, powerful and often antagonistic beings who represent chaos and primal forces.
5. Niflheim: A realm of mist and darkness, associated with cold and death, where the primordial well of Hvergelmir resides.

6.   Muspelheim: A realm of fire and heat, inhabited by fire giants and ruled by the fire god Surtr.

7.   Svartalfheim: The domain of the dark elves and dwarves, skilled craftsmen and miners who dwell in subterranean realms.

8.   Helheim: The underworld ruled by the goddess Hel, where those who die of illness or old age go after death.

9.   Nidavellir: Another realm inhabited by dwarves, known for its vast underground caverns and forges.

Each of these realms has its own mythology, inhabitants, and significance within the broader framework of Norse cosmology. The interconnectedness of these worlds, facilitated by Yggdrasil, underscores the dynamic and interdependent nature of the Norse universe.

The concept of multiple worlds in Norse mythology reflects humanity's quest for understanding the cosmos and the diverse manifestations of existence. These myths provide a framework for exploring themes of creation, cosmogony, and the interactions between gods, humans, and other beings. They serve as a testament to the rich imagination and storytelling traditions of ancient cultures, offering insights into their beliefs, values, and worldview.

Plato and the Realm of Forms: In ancient Greece, Plato proposed the existence of an abstract Realm of Forms, where every concept and object in our world has a perfect counterpart. This idea, while metaphysical, parallels modern notions of other dimensions where different rules may apply.

- Plato's concept of the Realm of Forms, also known as the Theory of Forms or Ideas, is one of the most influential and enduring philosophical ideas in Western thought. Originating in ancient Greece, this metaphysical framework posits the existence of a transcendent realm beyond the physical world, where immutable, perfect forms or ideas exist as the true essence of reality.

According to Plato, the physical world that we perceive through our senses is merely a shadowy reflection or imperfect instantiation of these ideal forms. Every object, concept, or quality in the physical realm, such as beauty, justice, or the form of a horse, is derived from its corresponding perfect form in the Realm of Forms. These forms are eternal, unchanging, and absolute, serving as the ultimate reality upon which the material world is based.

Plato's allegory of the cave illustrates this concept vividly. In the allegory, individuals are depicted as prisoners chained within a dark cave, facing a wall where shadows cast by a fire behind them are the only reality they perceive. These shadows represent the imperfect reflections of the true forms, which the prisoners cannot directly see. Only through philosophical contemplation and the ascent out of the cave into the light of the sun, symbolizing the realm of forms, can one attain true knowledge and understanding of reality.

While Plato's theory of Forms is primarily metaphysical and speculative, it bears striking similarities to modern scientific and philosophical concepts, particularly those concerning higher dimensions and alternate realities. Modern physics, for example, postulates the existence of dimensions beyond the familiar three spatial dimensions and one temporal dimension we perceive. String theory, for instance, proposes the existence of multiple dimensions, each with its own set of physical laws and properties.

Similarly, the concept of alternate realities or parallel universes, often explored in science fiction and speculative physics, shares parallels with Plato's Realm of Forms. These alternate realities may have different fundamental constants, laws of nature, or even entirely different forms of existence, mirroring Plato's idea of a transcendent realm where different rules apply.

While Plato's Realm of Forms remains a philosophical abstraction, it continues to provoke thought and inspire inquiry into the nature of reality, the limitations of human perception, and the possibility of realms beyond our immediate experience. Whether viewed through the lens of ancient philosophy or modern science, the notion of transcendent forms or alternate dimensions offers a tantalizing glimpse into the mysteries of existence and the potential richness of reality beyond what we can perceive.

The Hindu Multiverse: Hindu cosmology is perhaps one of the earliest examples of a multiverse, with scriptures mentioning an infinite number of universes, each with its own gods and beings, existing and reoccurring through cosmic cycles of creation and destruction.

- Hindu cosmology offers a profound and expansive vision of reality, encompassing not just a single universe but an infinite multiverse teeming with diverse worlds, gods, and beings. This cosmological worldview, rooted in ancient Hindu scriptures such as the Vedas, Puranas, and Upanishads, portrays a dynamic and cyclic universe where creation and destruction unfold across vast cosmic cycles.

  Central to Hindu cosmology is the concept of "Loka," which denotes realms or worlds within the multiverse. These Lokas are inhabited by various deities, celestial beings, demons, humans, animals, and other entities, each playing a distinct role in the cosmic drama of existence. The Lokas are not confined to physical dimensions alone but encompass different planes of reality, ranging from the material to the spiritual.

One of the fundamental principles of Hindu cosmology is the idea of cyclical creation and destruction, known as "Srishti" and "Pralaya," respectively. According to Hindu cosmology, the universe undergoes an infinite series of these cycles, each spanning immense durations of time known as "Kalpas." During the process of creation, the Supreme Being, often referred to as Brahman or Ishvara, manifests the universe from the cosmic void through a divine act of creation. This universe is but one among countless others, each emerging from and returning to the cosmic ocean of existence.

Within this vast multiverse, different Lokas exist on different planes of reality, each governed by its own set of laws and inhabited by its own pantheon of deities. The highest realm is "Satya Loka" or "Brahma Loka," the abode of Lord Brahma, the creator deity, where timeless wisdom and ultimate truth reside. Beneath this lie the celestial realms, such as "Swarga Loka" or "Devaloka," inhabited by benevolent deities and celestial beings who enjoy immense power and bliss. On the earthly plane, known as "Bhuloka," human beings reside, along with other terrestrial beings. Below Bhuloka are the realms of the underworld, including "Patala Loka," inhabited by demons and other malevolent entities.

The Hindu multiverse is not just a static framework but a dynamic and interconnected web of existence, where beings traverse different Lokas through the cycles of birth, death, and rebirth, known as "Samsara." This cyclical nature of existence is governed by the law of karma, which dictates the consequences of one's actions and determines one's future births and experiences.

Hindu cosmology offers a rich tapestry of myths, legends, and philosophical insights into the nature of reality and the mysteries of existence. It invites contemplation on the interconnectedness of all life, the impermanence of the physical world, and the eternal quest for spiritual realization. Through its portrayal of the multiverse, Hindu cosmology inspires awe and wonder at the vastness and complexity of the cosmos, inviting seekers to explore the depths of consciousness and the ultimate nature of reality.

Medieval and Renaissance Views

The Kabbalah and Jewish Mysticism: The Kabbalah, a form of Jewish mysticism, describes the Ein Sof (Infinite) and the process of creation through which multiple worlds are formed. This spiritual and mystical interpretation suggests layers of reality beyond our immediate perception.

- The Kabbalah, a mystical tradition within Judaism, offers a profound and intricate cosmology that delves into the nature of existence, the divine, and the interconnectedness of all things. Central to Kabbalistic thought is the concept of Ein Sof, often translated as the "Infinite" or the "Endless." Ein Sof represents the transcendent, ineffable aspect of the divine that is beyond human comprehension and description. It is the ultimate source and essence from which all existence emanates.

  According to Kabbalistic teachings, the process through which the divine manifests and creates the universe is known as "Sefirot." The Sefirot are divine emanations or attributes through which Ein Sof reveals itself and interacts with creation. These emanations form a hierarchical structure, often depicted as a tree known as the "Tree of Life." Each Sefira represents a different aspect of the divine and plays a distinct role in the ongoing process of creation and sustenance.

Within the Kabbalistic framework, the process of cre-ation is multifaceted and unfolds through multiple stages or worlds, known as "Olamot." These worlds are not physical realms but rather spiritual dimensions or levels of reality that exist beyond our immediate perception. The Kabbalah describes four primary worlds:

1. **Atzilut:** This is the world of emanation, the high-est and most divine realm, where the pure essence of the divine is fully revealed. Atzilut is characterized by unity, harmony, and the direct presence of the divine.
2. **Beriah:** Also known as the world of creation, Beriah is where the initial stages of creation take place. It is the realm of intellect and archetypal forms, where the divine ideas or blueprints for the universe exist.
3. **Yetzirah:** The world of formation, Yetzirah is where the divine energies coalesce and take on more concrete forms. It is associated with the realm of angels, spiritual forces, and the shaping of cosmic structures.
4. **Assiyah:** This is the world of action, the lowest and most material realm. Assiyah is where physical reality, including the material universe and human existence, manifests and unfolds.

Each of these worlds represents a different level of spir-itual reality, with its own unique characteristics and inhabi-tants. The process of creation involves the descent of divine energy and consciousness through these worlds, culminating in the manifestation of the physical universe.

The Kabbalistic understanding of the multilayered nature of reality suggests that there are deeper dimensions beyond the physical realm that are accessible through spiritual prac-tice and contemplation. By exploring the mysteries of the di-vine and the structure of creation, practitioners of Kabbalah seek to deepen their understanding of the cosmos and their relationship to the divine source.

In summary, the Kabbalah offers a rich and intricate cosmology that transcends the boundaries of conventional knowledge and perception. Through its teachings on Ein Sof, Sefirot, and Olamot, Kabbalistic mysticism invites seekers to explore the hidden dimensions of reality and to contemplate the divine mysteries that lie beyond our immediate grasp.

Giordano Bruno's Infinite Universe: In the Renaissance, the philosopher Giordano Bruno was one of the first Europeans to propose that the stars are distant suns surrounded by their own exoplanets, potentially hosting life, suggesting a universe much larger and more diverse than previously thought.

- Giordano Bruno, a Renaissance philosopher and visionary thinker, made significant contributions to the field of cosmology with his bold and revolutionary ideas about the universe. In an era dominated by religious dogma and Aristotelian cosmology, Bruno's notions challenged prevailing beliefs and laid the groundwork for modern conceptions of the cosmos.

  Central to Bruno's cosmology was the idea of an infinite universe teeming with countless stars, each potentially orbited by planets akin to our own. At a time when the prevailing view in Europe was that Earth lay at the center of a finite and geocentric universe, Bruno's vision was radical and transformative.

  Bruno's conception of the universe was heavily influenced by his philosophical and theological beliefs, which emphasized the interconnectedness and unity of all things. He rejected the idea of a hierarchical cosmos with Earth at its center, instead advocating for a universe without boundaries, where every point in space is potentially inhabited by life.

Bruno's cosmological vision was informed by his understanding of Copernican heliocentrism, which posited that Earth and other planets orbit the Sun, as well as by his study of the works of ancient philosophers such as Lucretius and Nicholas of Cusa. Drawing upon these sources, Bruno argued that the stars were distant suns similar to our own, each with its own system of planets and potentially harboring life.

Bruno's ideas about the plurality of worlds and the vastness of the cosmos were articulated in his seminal work, "De l'infinito, universo e mondi" ("On the Infinite Universe and Worlds"), published in 1584. In this work, Bruno challenged the prevailing geocentric model of the universe and proposed a radical vision of an infinite and diverse cosmos.

Unfortunately, Bruno's ideas were far ahead of his time, and his bold assertions brought him into conflict with the religious and intellectual authorities of his day. His advocacy for a heliocentric universe and his rejection of orthodox religious doctrines led to accusations of heresy, and he was eventually imprisoned and executed by the Roman Inquisition in 1600.

Despite the tragic end to his life, Giordano Bruno's contributions to cosmology and his vision of an infinite universe laid the groundwork for future generations of astronomers and philosophers. His ideas challenged entrenched beliefs and helped pave the way for the scientific revolution of the seventeenth century, which transformed our understanding of the cosmos and our place within it. Today, Bruno is remembered as a pioneering thinker whose bold ideas continue to inspire exploration and inquiry into the nature of the universe.

Scientific Revolution to 19th Century

Newton and the Mechanical Universe: The enlightenment brought a more mechanical view of the universe. Although Isaac Newton's laws of motion and universal gravitation didn't explicitly suggest parallel worlds, they opened the door to thinking about the universe in terms of laws that could be universally applied, possibly beyond our own cosmos.

- The Enlightenment period marked a significant shift in humanity's understanding of the natural world, characterized by a growing emphasis on reason, empiricism, and the scientific method. One of the key developments during this era was the emergence of a more mechanical view of the universe, which sought to explain natural phenomena in terms of mathematical laws and mechanical principles. At the forefront of this transformation was the groundbreaking work of Sir Isaac Newton, whose laws of motion and universal gravitation revolutionized our understanding of the cosmos.

  Newton's laws of motion, as articulated in his seminal work "Philosophiæ Naturalis Principia Mathematica" (Mathematical Principles of Natural Philosophy) published in 1687, laid the foundation for classical mechanics. These laws provided a comprehensive framework for understanding the motion of objects and the forces acting upon them. Newton's first law, often referred to as the law of inertia, states that an object will remain at rest or in uniform motion unless acted upon by an external force. His second law relates the force acting on an object to its mass and acceleration, while his third law states that for every action, there is an equal and opposite reaction.

  In addition to his laws of motion, Newton formulated the law of universal gravitation, which describes the force of attraction between two objects with mass. According to this law, every particle of matter in the universe attracts every other particle with a force that is directly proportional to the product of their masses and inversely proportional to the square of the distance between them. Newton's law of universal gravitation provided a unified explanation for celestial motion, including the orbits of planets around the Sun and the motion of the Moon around Earth.

Newton's mechanical worldview, grounded in mathematical rigor and empirical observation, transformed the way in which scientists and philosophers conceptualized the universe. Instead of viewing the cosmos as a realm governed by divine will or supernatural forces, Newtonian physics portrayed the universe as a vast, orderly machine operating according to deterministic laws.

While Newton's laws of motion and gravitation did not explicitly suggest the existence of parallel worlds or alternate dimensions, they did pave the way for a more expansive and universal conception of physical laws. By demonstrating that the same laws of motion and gravitation that govern objects on Earth also apply to celestial bodies throughout the cosmos, Newton's work suggested the possibility of a unified understanding of the universe that transcended the confines of our own solar system.

In this way, Newton's mechanical view of the universe opened the door to thinking about the cosmos in terms of laws that could be universally applied, potentially extending beyond our immediate surroundings. While Newton himself may not have speculated about parallel worlds, his contributions laid the groundwork for future generations of scientists and thinkers to explore the possibility of a broader, more interconnected reality governed by fundamental laws of nature.

The Romantic Perspective: During the Romantic period, literature and philosophy entertained the notion of vast, unknowable universes as a counterpoint to the Enlightenment's rationalism. Writers like Edgar Allan Poe in his work "Eureka" speculated about a plurality of worlds and even a sort of proto-multiverse.

- The Romantic period, spanning roughly from the late 18th to the mid-19th century, was characterized by a profound shift in artistic, literary, and philosophical sensibilities. Reacting against the rationalism and mechanistic worldview of the Enlightenment, Romantic thinkers embraced emotion, intuition, and imagination as powerful sources of truth and meaning. Central to the Romantic perspective was a fascination with the mysterious, the sublime, and the infinite—a departure from the Enlightenment's emphasis on reason and empiricism.

  In the realm of literature and philosophy, Romantic thinkers explored themes of cosmic vastness, the unknown, and the ineffable nature of existence. One prominent example of this exploration is found in the work of Edgar Allan Poe, particularly his essay "Eureka: A Prose Poem," published in 1848. In "Eureka," Poe ventures into the realms of cosmology and metaphysics, offering his own speculative theories about the nature of the universe and humanity's place within it.

  Poe's "Eureka" is a bold and ambitious work that defies easy categorization. Part philosophical treatise, part poetic meditation, "Eureka" explores themes of cosmology, ontology, and the search for ultimate truth. Drawing on scientific knowledge of his time as well as his own imaginative insights, Poe proposes a theory of the universe that anticipates many concepts of modern cosmology, including the idea of a dynamic and evolving cosmos.

Central to Poe's cosmology is the notion of a plurality of worlds—a concept that resonates with earlier speculative thinkers such as Giordano Bruno. Poe envisions a universe teeming with innumerable galaxies, stars, and planets, each inhabited by intelligent beings like ourselves. He speculates about the possibility of other worlds existing beyond our own, stretching into infinity—a vision that anticipates the modern concept of the multiverse.

Poe's vision of the cosmos in "Eureka" is characterized by its grandeur, its mystery, and its sense of awe-inspiring beauty. He describes the universe as a vast and intricate web of interconnected forces, governed by universal laws that transcend human understanding. Poe's prose is infused with a sense of wonder and reverence for the sublime, echoing the Romantic fascination with the sublime as a source of inspiration and transcendence.

While Poe's speculations in "Eureka" may lack the empirical rigor of scientific cosmology, they nevertheless capture the spirit of Romanticism—the quest for transcendence, the celebration of the imagination, and the recognition of the infinite complexity of the universe. Through his exploration of the cosmos, Poe invites readers to contemplate the mysteries of existence and to confront the profound questions that lie at the heart of human experience.

In summary, the Romantic perspective during the 18th and 19th centuries represented a departure from the rationalism of the Enlightenment, embracing themes of cosmic vastness, the unknown, and the sublime. Writers like Edgar Allan Poe, in works such as "Eureka," speculated about the nature of the universe and humanity's place within it, offering imaginative visions of a cosmos that transcends the limitations of human understanding.

Modern Philosophical Thought

Existentialism and Parallel Lives: In the 20th century, existential philosophers like Jean-Paul Sartre and existentialist writers such as Franz Kafka explored themes of alienation, freedom, and the multiplicity of choices, indirectly suggesting the weight of 'other lives' we might have led under different circumstances.

- Existentialism, a philosophical movement that emerged in the 20th century, focused on the individual's struggle to find meaning and authenticity in a seemingly absurd and indifferent universe. Central to existential thought were themes of freedom, choice, and the experience of being alone in an indifferent world. Existential philosophers such as Jean-Paul Sartre, Albert Camus, and existentialist writers like Franz Kafka, explored these themes through their writings, often delving into the notion of parallel lives and the myriad possibilities that exist within the human experience.

  Jean-Paul Sartre, one of the key figures of existentialism, famously articulated the concept of "radical freedom" in his work "Being and Nothingness." According to Sartre, human beings are condemned to be free, meaning that we are fundamentally responsible for our choices and actions in a universe devoid of inherent meaning or purpose. This existential freedom entails the burden of creating our own values and meanings in a world characterized by uncertainty and contingency.

  Sartre's exploration of freedom and choice inevitably leads to the consideration of alternate possibilities and parallel lives. In his play "No Exit," for example, Sartre presents a vision of hell where the characters are trapped together for eternity, forced to confront the consequences of their actions and decisions. The play serves as a metaphor for the existential condition, highlighting the ways in which individuals are bound by their choices and their relationships with others.

Similarly, the writings of Franz Kafka often evoke a sense of existential angst and alienation, portraying characters who grapple with the absurdity and futility of existence. In works such as "The Trial" and "The Metamorphosis," Kafka explores themes of isolation, guilt, and the inherent absurdity of human life. His characters often find themselves trapped in nightmarish situations, unable to escape the existential dilemmas that confront them.

Through their explorations of freedom, choice, and the absurdity of existence, existential philosophers and writers indirectly suggest the existence of parallel lives and the myriad possibilities that lie within the human experience. While we may never know the paths not taken or the lives not lived, existentialism invites us to confront the reality of our own existence and to embrace the uncertainty and contingency that define the human condition.

In this sense, existentialism offers a provocative lens through which to consider the weight of 'other lives' we might have led under different circumstances. By emphasizing the individual's responsibility for their choices and actions, existentialism underscores the significance of each moment and decision in shaping the course of our lives. Ultimately, existentialism invites us to confront the paradox of freedom and to embrace the possibilities that arise from the inherent uncertainty of human existence.

Quantum Mechanics and Philosophy: The advent of quantum mechanics revived philosophical interest in parallel worlds. Philosophers of science debated the implications of quantum phenomena like superposition and entanglement, questioning how these might influence our understanding of reality.

- The advent of quantum mechanics in the early 20th century revolutionized our understanding of the fundamental nature of reality, introducing concepts that challenged classical notions of determinism, causality, and the very fabric of space and time. Quantum phenomena such as superposition, entanglement, and wave-particle duality opened new avenues for philosophical inquiry, prompting intense debate among philosophers of science about the nature of reality and the implications of quantum theory for our understanding of the universe.

  One of the most intriguing aspects of quantum mechanics from a philosophical perspective is the concept of superposition. According to quantum theory, particles such as electrons can exist in multiple states simultaneously, a phenomenon known as superposition. This challenges our intuitive understanding of reality, which is based on classical physics, where objects are assumed to have definite properties at all times. Philosophers grappled with the question of what it means for a particle to be in a superposition of states and how this relates to our perception of reality.

  Entanglement, another key concept in quantum mechanics, further complicates our understanding of reality. When two particles become entangled, their properties become correlated in such a way that the state of one particle instantaneously influences the state of the other, regardless of the distance between them. This phenomenon, famously described by Albert Einstein as "spooky action at a distance," challenges classical notions of locality and separability, raising profound questions about the nature of causality and the interconnectedness of the universe.

The implications of these quantum phenomena for our understanding of reality sparked intense philosophical debates. Some philosophers argued that the indeterminacy inherent in quantum mechanics undermines the notion of a deterministic universe governed by fixed laws, opening the door to the possibility of genuine randomness and free will. Others questioned whether the apparent non-locality of entanglement could be reconciled with our conventional understanding of space and time, suggesting the need for a radical rethinking of the nature of reality.

These debates led to the development of various interpretations of quantum mechanics, each offering a different philosophical perspective on the meaning and implications of quantum theory. The Copenhagen interpretation, proposed by Niels Bohr and Werner Heisenberg, emphasizes the role of observation in determining the outcome of quantum experiments, highlighting the inherent subjectivity of measurement in quantum mechanics. The Many-Worlds interpretation, put forward by Hugh Everett III, suggests that every possible outcome of a quantum measurement corresponds to a separate branch of reality, leading to the existence of parallel worlds.

The philosophical implications of quantum mechanics continue to be a source of fascination and debate among philosophers of science. As our understanding of quantum phenomena deepens and our ability to manipulate and control quantum systems improves, the philosophical questions raised by quantum theory will remain at the forefront of scientific inquiry, challenging our assumptions about the nature of reality and our place within it.

Conclusion

Our historical journey reveals that the idea of parallel universes is not merely a product of contemporary science but a concept deeply rooted in human culture and intellectual history. These ideas have evolved from mythological and religious contexts to philosophical and scientific frameworks, showing a perennial fascination with the unknown. Each cultural and historical perspective offers unique insights into how humans understand their universe and, by extension, how we might continue to explore the vast possibilities suggested by the multiverse theory. As we turn towards more recent scientific developments in the next chapters, we carry with us a rich tapestry of human thought—a legacy that enriches our quest to understand the nature of reality and the many worlds it might contain.

~ 4 ~

THE SCIENCE OF PARALLEL UNIVERSES

In this chapter, we delve deeper into the realm of theoretical physics to explore the fascinating connections between string theory, cosmic inflation, and the scientific basis for the existence of parallel universes. As we unpack these complex ideas, we aim to provide a clear understanding of how contemporary science supports the possibility of multiple universes existing alongside our own.

String Theory: Vibrating Strings and Hidden Dimensions

String theory, a cornerstone of modern theoretical physics, proposes that the fundamental constituents of reality are not point-like particles, but rather tiny, vibrating strings. These strings operate in a landscape of ten or more dimensions—far beyond the three spatial dimensions and one time dimension that we experience. Let's explore how this theory contributes to our understanding of parallel universes.

Vibrating Strings: Each string in string theory vibrates at a specific frequency, and these vibrations correspond to different particles. The mass and charge of a particle are determined by the way the string vibrates.

- In the realm of string theory, one of the fundamental ideas is that elementary particles are not point-like objects but rather tiny, one-dimensional "strings" that vibrate at different frequencies. These strings can oscillate in various modes, and the different modes of vibration correspond to different particle properties such as mass, charge, and spin.

  Each type of elementary particle in the Standard Model of particle physics is associated with a particular vibrational mode of the string. For example, a photon, the carrier of electromagnetic force, corresponds to a string vibrating in one particular mode, while an electron corresponds to a different mode of vibration. The vibrational pattern of the string determines the particle's properties, including its mass and charge.

  The concept of vibrating strings in string theory provides a unified framework for understanding the fundamental forces and particles of nature. By postulating that all particles are manifestations of vibrating strings, string theory aims to reconcile the principles of quantum mechanics with those of general relativity, thus providing a unified description of the fundamental forces of nature.

  Moreover, string theory suggests that the universe may have more dimensions than the familiar three spatial dimensions and one time dimension. In addition to these four dimensions, string theory posits the existence of extra spatial dimensions, which are compactified or "curled up" at extremely small scales. The vibrations of strings in these extra dimensions can give rise to different types of particles and interactions, leading to the rich diversity of phenomena observed in the universe.

While string theory remains a highly speculative and mathematically complex theory, it offers a promising framework for addressing some of the most profound questions in theoretical physics, such as the nature of spacetime, the origin of particle masses, and the unification of fundamental forces. The idea of vibrating strings provides a compelling and elegant picture of the fundamental constituents of matter and the forces that govern their interactions, paving the way for deeper insights into the nature of reality.

Extra Dimensions: The extra dimensions proposed by string theory are typically compactified, meaning they are curled up so tightly that they are invisible to us at macroscopic scales. Different ways of compactifying these dimensions lead to different physical laws, suggesting that other universes could exist with laws vastly different from our own.

- In the context of string theory, the notion of extra dimensions plays a crucial role in providing a unified framework for understanding the fundamental forces and particles of nature. Unlike the familiar three spatial dimensions (length, width, and height) and one time dimension that we experience in our everyday lives, string theory posits the existence of additional spatial dimensions beyond these four dimensions.

  The idea of extra dimensions arises naturally within the mathematics of string theory, which describes elementary particles as one-dimensional strings rather than point-like particles. In order for the mathematical consistency of string theory to hold, the theory requires a total of ten dimensions, with six additional spatial dimensions beyond the familiar three.

However, unlike the three large spatial dimensions that we perceive, the extra dimensions proposed by string theory are typically compactified or "curled up" at extremely small scales. These compact dimensions are so tiny and tightly wound that they are effectively invisible to us at macroscopic scales. In other words, while we cannot directly observe or detect these extra dimensions with our current technology, their existence is inferred from the mathematical formalism of string theory.

The concept of compactified extra dimensions has profound implications for our understanding of the universe. Different ways of compactifying these extra dimensions can lead to different geometries and topologies, which in turn give rise to different physical laws and phenomena in the four-dimensional spacetime that we inhabit. This suggests that there could be multiple possible configurations of the extra dimensions, each corresponding to a different "vacuum state" or set of physical laws.

In this way, string theory provides a framework for the existence of a "landscape" of possible universes, each with its own unique set of physical constants, particle masses, and fundamental forces. These other universes could have laws of physics that are vastly different from our own, leading to the possibility of entirely different forms of matter, energy, and even dimensions of spacetime.

While the concept of extra dimensions and the multiverse remains speculative, it offers a tantalizing glimpse into the potential richness and diversity of the cosmos beyond what we can directly observe. Exploring the implications of extra dimensions within the framework of string theory continues to be an active area of research in theoretical physics, with the hope of uncovering new insights into the nature of reality and our place within it.

The Landscape of String Theory: This theory introduces the concept of a "landscape" of possible universes. Each point in this landscape represents a different way the extra dimensions could be compactified, leading to a different set of physical laws. Thus, string theory allows for a vast number of possible universes in the multiverse.

- In the landscape of string theory, each point represents a distinct configuration of the extra dimensions, resulting in a unique set of physical laws and properties. This concept, known as the "landscape," arises from the inherent flexibility of string theory, which allows for a multitude of ways in which the extra dimensions can be compactified. The landscape theory suggests that our universe is just one of many possible universes that could exist within the broader framework of string theory, each with its own distinct characteristics.

  The idea of the landscape emerges from the mathematical formalism of string theory, which describes elementary particles as one-dimensional strings vibrating in higher-dimensional spacetime. In addition to the familiar three spatial dimensions and one time dimension, string theory requires the existence of extra spatial dimensions, typically six or seven in total, in order for the theory to be mathematically consistent.

  However, the precise geometry and topology of these extra dimensions are not determined by the fundamental equations of string theory. Instead, there are a vast number of possible ways in which the extra dimensions can be compactified, each corresponding to a different vacuum state or set of physical laws. These different configurations of the extra dimensions give rise to different universes within the landscape of string theory.

The landscape of string theory has profound implications for our understanding of the cosmos. It suggests that our universe is just one of many possible realizations of the laws of physics, with the properties of our universe being determined by the specific configuration of the extra dimensions. Moreover, the landscape theory implies that there could be universes with vastly different physical laws and properties from our own, leading to the possibility of exotic phenomena and forms of matter that are not observed in our universe.

The landscape of string theory has sparked intense debate within the scientific community. Critics argue that the concept of the landscape is inherently untestable and therefore not scientifically valid, as it does not make specific predictions that can be verified through observation or experiment. Proponents, however, contend that the landscape theory provides a natural explanation for the apparent fine-tuning of the laws of physics in our universe, suggesting that our universe is just one of many possible outcomes of the underlying dynamics of string theory.

Regardless of these debates, the landscape of string theory represents a provocative and fertile ground for exploring the nature of reality and the possibilities inherent in the framework of string theory. It challenges our preconceptions about the uniqueness of our universe and invites us to consider the vast diversity of possible universes that could exist within the multiverse.

Cosmic Inflation and the Multiverse

Cosmic inflation, a theory proposed by physicist Alan Guth in the early 1980s, describes a period of extremely rapid expansion of the universe just after the Big Bang. Inflation not only explains the uniformity of the cosmic microwave background radiation but also has implications for the multiverse.

Eternal Inflation: In some models, inflation never completely stops. It ceases in some regions, forming what we perceive as our universe, but continues elsewhere indefinitely. This ongoing process can lead to the creation of many isolated "bubble universes," each potentially governed by different physical laws.

- Eternal inflation is a fascinating concept in cosmology that arises from certain models of cosmic inflation, a theory proposed to explain the large-scale structure and uniformity of the universe. According to these models, inflation is a period of extremely rapid expansion that occurred in the early universe, shortly after the Big Bang. While inflation eventually ceases in some regions, leading to the formation of observable universes like our own, it continues indefinitely in other regions, giving rise to a process of eternal expansion.

  In the context of eternal inflation, the universe is envisioned as a vast and ever-expanding "multiverse," composed of many isolated regions or "pocket universes." Each of these pocket universes represents a separate bubble of spacetime, with its own distinct properties, laws of physics, and initial conditions. These bubbles of spacetime are continuously being created and expanding, leading to a proliferation of universes within the multiverse.

  One of the key features of eternal inflation is that it leads to the spontaneous formation of bubble universes within the larger multiverse. As inflation continues in certain regions of spacetime, quantum fluctuations can cause the energy density to fluctuate, leading to the spontaneous creation of regions where inflation ceases and a new bubble universe is born. These bubble universes are separated from each other by regions of ongoing inflation, making communication between them effectively impossible.

The concept of eternal inflation has profound implications for our understanding of the cosmos. It suggests that our universe is just one of many possible outcomes of the underlying dynamics of inflation, with the properties of our universe being determined by the specific conditions that prevailed during its formation. Moreover, the existence of a multiverse populated by a vast number of bubble universes raises the possibility that there could be a diversity of physical laws and phenomena beyond what is observed in our own universe.

While the concept of eternal inflation remains speculative and subject to ongoing research and debate, it offers a compelling framework for addressing some of the most profound questions in cosmology. By positing the existence of a multiverse populated by an infinite number of bubble universes, eternal inflation challenges our preconceptions about the uniqueness of our universe and invites us to consider the vastness and complexity of the cosmos beyond what is observable to us. As scientists continue to explore the implications of eternal inflation, it promises to shed new light on the nature of reality and our place within the multiverse.

Bubble Universes: These universes are separated from each other by expanding space and may never interact. However, each bubble universe can have different properties, depending on how inflation ended in that region. This variation across universes is a key aspect of the multiverse theory supported by inflation.

- Bubble universes are hypothetical regions of spacetime that exist within the framework of certain cosmological theories, particularly those involving cosmic inflation and the multiverse hypothesis. In these theories, bubble universes arise as a consequence of the dynamics of inflation, a period of rapid expansion that occurred in the early universe.

The concept of bubble universes is closely tied to the idea of eternal inflation, which posits that inflation never completely stops but continues indefinitely in certain regions of spacetime. As inflation proceeds, quantum fluctuations can cause the energy density to fluctuate, leading to the spontaneous formation of regions where inflation ceases and bubble universes are born.

Each bubble universe represents a separate and self-contained domain of spacetime, characterized by its own distinct properties, laws of physics, and initial conditions. These properties can vary from one bubble universe to another, depending on how inflation ended in that particular region. For example, the strength of fundamental forces, the values of physical constants, and the nature of matter and energy may differ between bubble universes.

The key feature of bubble universes is that they are typically separated from each other by expanding space and may never interact with one another. This separation arises from the ongoing process of inflation, which drives the expansion of spacetime between bubble universes faster than the speed of light. As a result, communication or travel between bubble universes is effectively impossible, making each bubble universe a self-contained entity with its own distinct history and evolution.

The variation of properties across bubble universes is a central aspect of the multiverse theory supported by inflation. According to this view, our universe is just one of many possible outcomes of the underlying dynamics of inflation, with the properties of our universe being determined by the specific conditions that prevailed during its formation. The existence of a multitude of bubble universes, each with its own set of physical laws and phenomena, reflects the vastness and complexity of the multiverse.

While the concept of bubble universes remains speculative and subject to ongoing research and debate, it offers a compelling framework for addressing some of the deepest questions in cosmology. By positing the existence of a multiverse populated by a vast number of bubble universes, inflationary cosmology challenges our conventional notions of cosmic uniqueness and invites us to consider the rich diversity of possible universes that could exist beyond our own.

Quantum Mechanics and Universe Branching

Building on the foundations laid in the previous chapters, we revisit quantum mechanics to understand its implications for the multiverse at a deeper level.

Quantum Decoherence: This process describes how quantum superpositions, such as those involving Schrödinger's Cat being both alive and dead, appear to "collapse" into one of the possible states. In the many-worlds interpretation, each possible outcome of a quantum event is realized in a separate branch of the universe.

- Quantum decoherence is a fundamental concept in quantum mechanics that describes the process by which quantum systems transition from a state of superposition to a state where one of the possible outcomes is observed. This phenomenon is crucial for understanding the apparent collapse of the quantum wavefunction and the emergence of classical behavior at the macroscopic scale.

  In quantum mechanics, particles such as electrons and photons can exist in multiple states simultaneously, a phenomenon known as superposition. One of the most famous illustrations of superposition is Schrödinger's Cat, a thought experiment proposed by physicist Erwin Schrödinger. In this scenario, a cat is placed in a box with a radioactive atom that has a 50% chance of decaying within a certain time frame. According to quantum mechanics, until the box is opened and the cat is observed, it exists in a superposition of being both alive and dead.

However, when the box is opened and the cat is observed, only one of the possible outcomes is observed—the cat is either alive or dead, not both. This apparent collapse of the wavefunction from a superposition to a single outcome is known as quantum decoherence.

In the many-worlds interpretation of quantum mechanics, proposed by physicist Hugh Everett III, quantum decoherence is understood differently. Instead of the wavefunction collapsing to a single outcome, the many-worlds interpretation suggests that all possible outcomes of a quantum event are realized in separate branches of the universe. In the case of Schrödinger's Cat, the universe would split into two branches —one where the cat is alive and another where the cat is dead —each representing a different outcome of the experiment.

According to the many-worlds interpretation, the apparent collapse of the wavefunction is an illusion created by our subjective experience of reality. In reality, all possible outcomes of a quantum event occur simultaneously in different branches of the multiverse, each branching off from the other at the moment of decoherence.

Quantum decoherence has profound implications for our understanding of quantum mechanics and the nature of reality. It provides a mechanism for reconciling the apparent randomness and indeterminacy of quantum events with the deterministic behavior observed at the macroscopic scale. Moreover, the many-worlds interpretation offers a radical reinterpretation of quantum mechanics, suggesting that the universe is constantly branching into multiple parallel realities, each representing a different outcome of quantum events.

While the many-worlds interpretation remains controversial and speculative, quantum decoherence is a well-established phenomenon that has been observed in numerous experiments. It plays a crucial role in bridging the gap between quantum mechanics and classical physics, providing insight into the transition from the microscopic to the macroscopic world. As our understanding of quantum decoherence continues to deepen, it promises to shed new light on the nature of reality and our place within the quantum multiverse.

Parallel Quantum Histories: According to the many-worlds interpretation, every quantum event generates a bifurcation of the universe, leading to a near-infinite number of parallel universes where everything that can happen does happen in some universe.

- The concept of parallel quantum histories stems from the many-worlds interpretation (MWI) of quantum mechanics, proposed by physicist Hugh Everett III in the 1950s. According to the MWI, every quantum event results in the branching of reality into multiple parallel universes, each representing a different outcome of the event. These parallel universes, or quantum histories, exist independently of each other and continue to evolve over time, branching further with each subsequent quantum event.

  In the framework of the MWI, the universe is viewed as a vast and constantly branching multiverse, where every possible outcome of a quantum event is realized in a separate parallel universe. For example, in the famous double-slit experiment, where a particle exhibits wave-like behavior when not observed, according to the MWI, the particle simultaneously goes through both slits in different branches of the multiverse, leading to an interference pattern on the screen.

The concept of parallel quantum histories challenges the traditional notion of a single, deterministic reality and instead suggests that reality is inherently probabilistic and constantly branching into multiple possibilities. In each parallel universe, all possible outcomes of a quantum event are realized, leading to a near-infinite number of parallel universes where everything that can happen does happen in some universe.

This idea has profound implications for our understanding of the nature of reality and the role of observers in quantum mechanics. According to the MWI, observers themselves are also subject to the branching of reality, with each observer experiencing a unique sequence of events based on their interactions with the quantum world. In this view, the apparent collapse of the wavefunction is an illusion created by the subjective experience of observers, who perceive only one of the many possible outcomes of a quantum event.

The concept of parallel quantum histories has sparked intense debate within the scientific community, with proponents arguing that it provides a simple and elegant explanation for the puzzling phenomena of quantum mechanics. Critics, however, question the empirical testability of the MWI and argue that it leads to an over-proliferation of parallel universes, making it difficult to distinguish between different interpretations of quantum mechanics.

Despite these challenges, the idea of parallel quantum histories continues to captivate the imagination of physicists and philosophers alike, offering a provocative vision of the quantum multiverse and the nature of reality itself. As our understanding of quantum mechanics deepens and our ability to probe the quantum world improves, the concept of parallel quantum histories promises to remain a fertile ground for exploration and inquiry into the fundamental nature of the universe.

Conclusion

The theories discussed in this chapter—string theory with its extra dimensions and diverse universe possibilities, cosmic inflation's generation of bubble universes, and quantum mechanics' implications for universe branching—each contribute to a compelling scientific framework supporting the existence of a multiverse. As we proceed, keep in mind that while these concepts stretch the imagination, they are rooted in rigorous scientific inquiry and remain the subject of active research and debate within the physics community.

In the next chapter, we will explore how these theoretical insights translate into the metaphysical and philosophical realms, challenging our perceptions and expanding our understanding of reality.

~ 5 ~

MULTIVERSE IN POP CULTURE

In this chapter, we explore how the concept of the multiverse has permeated popular culture, influencing literature, film, television, and other media. This cultural exploration helps us understand not only how the multiverse theory is represented in mainstream entertainment but also how these portrayals impact our understanding of and engagement with complex scientific and philosophical ideas.

The Multiverse in Literature

Early Influences: The idea of parallel universes has been a staple in science fiction literature for decades. Authors like H.G. Wells in "The Time Machine" hinted at alternate realities through the medium of time travel, setting a precedent for exploring 'what could have been' in different timelines.

- The concept of parallel universes, also known as alternate realities or alternate dimensions, has been a recurring theme in science fiction literature for well over a century. Authors have used parallel universes as a narrative device to explore intriguing questions about the nature of reality, the

49

possibilities of time travel, and the consequences of divergent choices. One of the earliest and most influential works of science fiction to explore the idea of parallel universes is H.G. Wells' "The Time Machine," published in 1895.

In "The Time Machine," Wells introduces readers to an unnamed Time Traveller who invents a machine capable of traversing through time. As the Time Traveller journeys into the distant future, he encounters a world inhabited by two distinct species—the gentle Eloi and the predatory Morlocks. However, it is not just the distant future that the Time Traveller explores; he also ventures into the distant past and witnesses the slow death of the Earth's sun.

While "The Time Machine" primarily focuses on time travel, it also hints at the concept of alternate realities and parallel universes. Through the Time Traveller's journeys, Wells suggests that different points in time represent different branches of reality, each with its own set of possibilities and consequences. By exploring these alternate timelines, Wells invites readers to ponder questions about fate, free will, and the interconnectedness of past, present, and future.

Wells' exploration of parallel universes in "The Time Machine" set a precedent for subsequent works of science fiction, inspiring generations of authors to further develop and expand upon this theme. In the decades that followed, parallel universes became a staple trope in science fiction literature, appearing in works by authors such as Philip K. Dick, Isaac Asimov, and Arthur C. Clarke.

For example, Philip K. Dick's novel "The Man in the High Castle" imagines a world in which the Axis Powers won World War II, leading to a radically different outcome for the course of history. Similarly, Isaac Asimov's "The End of Eternity" explores the consequences of altering the past through time travel, leading to the creation of divergent timelines and parallel universes.

In Arthur C. Clarke's "The Light of Other Days," co-authored with Stephen Baxter, the invention of a device capable of viewing events in the past allows humanity to glimpse into parallel universes and witness alternate histories. This exploration of parallel universes raises profound questions about the nature of reality and the implications of our choices on the course of history.

Overall, the idea of parallel universes in science fiction literature reflects humanity's enduring fascination with the unknown and the possibility of worlds beyond our own. Through imaginative storytelling, authors like H.G. Wells have provided readers with a glimpse into the myriad possibilities of the multiverse, inspiring us to contemplate the mysteries of existence and our place within the cosmos.

Modern Sci-Fi and Fantasy: Contemporary authors such as Neal Stephenson in "Anathem" or Philip Pullman in "His Dark Materials" series delve deeply into the concept of parallel worlds. These narratives often explore the ethical and moral implications of actions taken in alternate realities, pushing readers to think about the consequences of their decisions in a broader, more universal context.

- In contemporary science fiction and fantasy literature, authors like Neal Stephenson and Philip Pullman have continued to explore and expand upon the concept of parallel worlds, offering readers rich and immersive narratives that delve into the intricacies of alternate realities and their ethical and moral implications.

  Neal Stephenson's novel "Anathem" presents readers with a complex and intellectually stimulating exploration of parallel worlds within a vast and intricately imagined universe. Set in a world where scholars known as "avouts" live in secluded monasteries, "Anathem" follows the journey of a young mathematician named Erasmas as he discovers the existence of parallel universes and becomes embroiled in a cosmic struggle for the fate of humanity. Through Erasmas' experiences, Stephenson delves into profound questions about the nature of reality, the role of consciousness in shaping the universe, and the ethical dilemmas inherent in wielding knowledge and power.

Philip Pullman's "His Dark Materials" series takes readers on an epic journey across multiple parallel worlds, weaving together elements of fantasy, science fiction, and philosophy to create a rich and immersive narrative. Set in a world similar to our own but with subtle differences, the series follows the adventures of Lyra Belacqua as she discovers the existence of parallel universes and confronts powerful forces seeking to control them. Through Lyra's journey, Pullman explores themes of free will, destiny, and the nature of consciousness, challenging readers to consider the consequences of their actions in a multiverse where every choice has far-reaching implications.

Both "Anathem" and "His Dark Materials" push readers to think about the ethical and moral implications of actions taken in alternate realities, highlighting the interconnectedness of all things and the ripple effects that reverberate across the multiverse. By immersing readers in richly imagined worlds filled with complex characters and thought-provoking ideas, these narratives invite readers to contemplate their own place in the universe and the responsibilities that come with wielding knowledge and power.

In addition to "Anathem" and "His Dark Materials," contemporary authors across the science fiction and fantasy genres continue to explore the concept of parallel worlds in innovative and engaging ways. From exploring the consequences of time travel and alternate histories to grappling with the ethical dilemmas of creating and manipulating parallel universes, these narratives offer readers a captivating glimpse into the infinite possibilities of the multiverse and the profound questions that arise from contemplating the nature of reality.

The Multiverse in Film and Television

Blockbuster Films: The Marvel Cinematic Universe (MCU), with movies like "Doctor Strange" and "Avengers: Endgame," has brought the multiverse into the mainstream, presenting it as a central plot element that allows for an expansive storytelling landscape where characters can interact across different realities.

- The Marvel Cinematic Universe (MCU) has undeniably played a significant role in bringing complex and intriguing concepts like the multiverse into the mainstream consciousness through its blockbuster films. Films like "Doctor Strange" and "Avengers: Endgame" have prominently featured the multiverse as a central plot element, introducing audiences to the idea of parallel realities and the potential for interdimensional travel and interaction.

  "Doctor Strange," directed by Scott Derrickson, explores the mystical and metaphysical aspects of the Marvel Universe, introducing audiences to the concept of alternate dimensions and parallel realities. The film follows the journey of Dr. Stephen Strange, a brilliant but arrogant neurosurgeon who becomes a Master of the Mystic Arts after a career-ending accident. Through his training with the Ancient One, Strange learns to manipulate reality and traverse different dimensions, encountering strange and otherworldly beings along the way. The film's stunning visual effects and mind-bending sequences showcase the vastness and diversity of the multiverse, captivating audiences and expanding their understanding of the Marvel Universe beyond the confines of Earth.

"Avengers: Endgame," directed by Anthony and Joe Russo, further explores the multiverse concept within the context of the larger Marvel Cinematic Universe. The film follows the Avengers as they embark on a mission to reverse the devastating effects of Thanos' snap, which wiped out half of all life in the universe. Through the use of time travel and alternate realities, the Avengers encounter versions of themselves from different points in time and explore parallel universes where key events unfold differently. The film's ambitious narrative intertwines multiple timelines and dimensions, culminating in an epic battle that transcends the boundaries of space and time.

By incorporating the multiverse concept into its storytelling, the MCU has created an expansive narrative landscape where characters can interact across different realities, opening up endless possibilities for future storytelling. The introduction of the multiverse has allowed the MCU to explore new and exciting storylines, introduce new characters and concepts, and push the boundaries of what is possible within the Marvel Universe.

Moreover, the success of films like "Doctor Strange" and "Avengers: Endgame" has helped popularize the multiverse concept beyond the realm of comic books and into mainstream culture. Audiences have embraced the idea of parallel realities and interdimensional travel, fueling speculation and excitement for future MCU films and television series that explore the multiverse even further.

Overall, the MCU has played a pivotal role in bringing the multiverse into the mainstream, presenting it as a central plot element that enriches the storytelling experience and expands the Marvel Universe into new and exciting territory. As the MCU continues to evolve and grow, the multiverse promises to remain a key aspect of its narrative landscape, providing endless opportunities for creativity, exploration, and discovery.

Television Series: Shows like "Fringe" and "Rick and Morty" use the multiverse to explore themes of identity, destiny, and choice, providing viewers with a narrative tool to question the nature of reality and their place within it. These series often blend humor and drama, making the concept of the multiverse accessible and entertaining while still provoking thoughtful consideration.

- Television series have also embraced the concept of the multiverse, using it as a narrative device to explore complex themes and ideas while providing viewers with compelling storytelling. Shows like "Fringe" and "Rick and Morty" have become notable examples of how the multiverse can be used to delve into themes of identity, destiny, and choice, offering viewers a thought-provoking and often entertaining exploration of the nature of reality and their place within it.

  "Fringe," created by J.J. Abrams, Alex Kurtzman, and Roberto Orci, follows the exploits of the Fringe Division, a special task force tasked with investigating mysterious and often inexplicable events. As the series unfolds, it becomes increasingly apparent that the multiverse plays a central role in the show's mythology, with characters encountering alternate versions of themselves and exploring parallel realities where key events unfold differently. Through its exploration of the multiverse, "Fringe" delves into themes of identity and fate, raising questions about the nature of existence and the choices that define us.

Similarly, "Rick and Morty," created by Justin Roiland and Dan Harmon, uses the concept of the multiverse as a springboard for its irreverent and often surreal storytelling. The series follows the misadventures of an eccentric scientist named Rick Sanchez and his grandson Morty as they traverse the multiverse, encountering bizarre creatures, alternate dimensions, and existential crises along the way. Through its blend of humor and existential angst, "Rick and Morty" explores themes of free will, determinism, and the search for meaning in a vast and uncaring universe.

What sets both "Fringe" and "Rick and Morty" apart is their ability to use the concept of the multiverse to tell compelling stories that are both entertaining and thought-provoking. By blending elements of science fiction, drama, and humor, these series engage viewers on multiple levels, challenging them to question their assumptions about reality and the choices they make. Whether exploring the ethical implications of meddling with alternate realities or grappling with the consequences of one's actions, both shows offer viewers a captivating glimpse into the infinite possibilities of the multiverse.

Moreover, by using the multiverse as a narrative tool, "Fringe" and "Rick and Morty" make complex philosophical concepts accessible to a wide audience, fostering thoughtful consideration and discussion among viewers. Whether through its intricate mythology or its offbeat humor, each series invites viewers to contemplate their own place within the multiverse and the broader tapestry of existence, ultimately leaving them with a renewed appreciation for the mysteries of reality and the choices that shape our lives.

Impact on Public Understanding

Democratizing Complex Ideas: Pop culture has played a crucial role in democratizing complex scientific concepts, making them accessible to a broader audience. By embedding the idea of the multiverse in popular media, these concepts are no longer confined to academic discussions but are part of everyday conversations.

- Pop culture has indeed played a vital role in democratizing complex scientific concepts, including the notion of the multiverse, by making them more accessible and relatable to a broader audience. Through popular media such as movies, television shows, books, and video games, complex ideas are presented in engaging and entertaining ways, allowing audiences of all backgrounds to explore and understand them.

  One of the significant ways in which pop culture has democratized the concept of the multiverse is by incorporating it into mainstream narratives. Movies like "Doctor Strange," "Avengers: Endgame," and television series like "Fringe" and "Rick and Morty" have introduced the multiverse as a central plot element, presenting it in a visually compelling and relatable manner. By weaving the concept of the multiverse into stories that resonate with audiences, pop culture has made it more accessible and familiar, encouraging viewers to think critically about the nature of reality and the possibilities of alternate dimensions.

  Furthermore, pop culture has the power to simplify complex scientific concepts without sacrificing accuracy or depth. Through visual storytelling, creative narratives, and engaging characters, pop culture can distill intricate ideas into digestible and entertaining content that captivates audiences. Whether through captivating visuals, thought-provoking dialogue, or dramatic plot twists, pop culture presents the multiverse in ways that are engaging and comprehensible, sparking curiosity and encouraging further exploration.

Another way in which pop culture democratizes complex ideas is by fostering a sense of inclusivity and community. By presenting scientific concepts in a format that is accessible to a broad audience, pop culture creates opportunities for people from diverse backgrounds to engage with and discuss these ideas. Whether through online forums, social media platforms, or fan communities, pop culture enthusiasts can come together to share their thoughts, interpretations, and questions about the multiverse, fostering a sense of shared understanding and curiosity.

Moreover, by embedding complex scientific concepts like the multiverse in popular media, pop culture has the potential to inspire the next generation of scientists, innovators, and thinkers. For young audiences, seeing these ideas portrayed in movies, television shows, and other forms of entertainment can spark an interest in science and encourage them to explore these concepts further through education and research.

Overall, pop culture's role in democratizing complex scientific concepts like the multiverse cannot be overstated. By presenting these ideas in accessible, engaging, and relatable ways, pop culture has made them part of everyday conversations, inviting people of all backgrounds to explore, discuss, and appreciate the wonders of the universe. Through its power to inspire curiosity, foster inclusivity, and ignite imagination, pop culture continues to be a driving force in making complex ideas accessible to everyone.

Influence on Scientific Curiosity: Popular media has also sparked an increased interest in theoretical physics among the general public. This curiosity often leads to a deeper engagement with science, with more people tuning in to documentaries, attending lectures, and reading books that explore these topics in a serious academic context.

- Popular media's portrayal of complex scientific concepts, including theoretical physics topics like the multiverse, has indeed sparked a significant increase in scientific curiosity among the general public. Through movies, television shows, books, and other forms of entertainment, audiences are exposed to fascinating ideas that pique their interest and encourage them to learn more about the underlying science.

One of the key ways in which popular media influences scientific curiosity is by presenting scientific concepts in an accessible and engaging manner. By weaving these ideas into compelling narratives, popular media captures the imagination of audiences and ignites their curiosity about the mysteries of the universe. For example, movies like "Interstellar" and "The Theory of Everything" offer captivating glimpses into the world of theoretical physics, inspiring viewers to delve deeper into these topics and explore the underlying science.

Furthermore, popular media has the power to humanize scientists and make their work more relatable to the general public. Through movies, television shows, and documentaries, audiences are introduced to real-life scientists who are passionate about their work and dedicated to unraveling the secrets of the universe. By showcasing the human side of science and highlighting the personal journeys of scientists, popular media fosters a sense of connection and empathy, encouraging viewers to see themselves as part of the scientific community.

In addition to sparking initial interest, popular media also plays a crucial role in facilitating deeper engagement with scientific topics. Many people who are inspired by movies, television shows, or books that explore theoretical physics may seek out additional resources to learn more about these concepts in a serious academic context. This can lead to increased participation in science-related activities, such as attending lectures, joining discussion groups, or reading books written by experts in the field.

Moreover, popular media has the ability to reach a wide audience and spark conversations about scientific topics that might not otherwise be accessible to the general public. Through social media platforms, online forums, and fan communities, audiences can share their thoughts, questions, and insights about theoretical physics and engage in meaningful discussions with others who share their interests.

Overall, popular media's influence on scientific curiosity is undeniable. By presenting complex scientific concepts in accessible, engaging, and relatable ways, popular media inspires audiences to explore the wonders of the universe and encourages them to become more actively engaged with science. Whether through captivating narratives, inspiring portrayals of scientists, or thought-provoking discussions, popular media continues to play a vital role in fostering a deeper appreciation for theoretical physics and the pursuit of scientific knowledge.

Criticism and Misinterpretation

Scientific Accuracy: While the inclusion of the multiverse in pop culture has many benefits, it also comes with the risk of misinterpretation. Simplifications necessary for storytelling can sometimes lead to misunderstandings about the actual scientific theories.

• The portrayal of the multiverse in pop culture undoubtedly contributes to the accessibility and popularity of scientific concepts, but it also presents challenges in terms of scientific accuracy. While simplifications and creative liberties are often necessary for storytelling purposes, they can sometimes lead to misunderstandings or misinterpretations of the actual scientific theories behind the multiverse.

One of the primary challenges lies in balancing scientific accuracy with the demands of entertainment. Pop culture productions, such as movies, television shows, and novels, often prioritize narrative coherence, emotional engagement, and visual spectacle over strict adherence to scientific principles. As a result, certain aspects of the multiverse concept may be exaggerated, oversimplified, or misrepresented to fit the needs of the story.

For example, in some portrayals of the multiverse, alternate realities may be depicted as mirror images of our own, with only minor differences in the lives of characters or the course of history. While this makes for an easily understandable narrative device, it may not accurately reflect the complexity and diversity of theoretical models of the multiverse proposed by physicists.

Additionally, pop culture depictions of the multiverse may sometimes blur the lines between scientific theory and speculative fiction. Concepts like time travel, parallel dimensions, and alternate timelines are often conflated with the multiverse, leading to confusion about the distinct concepts and their implications within the realm of theoretical physics.

Moreover, the portrayal of the multiverse in pop culture may inadvertently perpetuate misconceptions or reinforce common tropes about science and scientists. Characters with extraordinary abilities or access to advanced technology may be presented as representing scientific knowledge, leading to unrealistic expectations or stereotypes about the capabilities of real-life scientists.

Despite these challenges, there are also opportunities for pop culture to contribute to scientific literacy and understanding. By engaging audiences with exciting and imaginative stories that feature scientific concepts like the multiverse, pop culture can inspire curiosity and encourage viewers to seek out more information about the underlying science.

To mitigate the risk of misinterpretation, creators of pop culture content can strive to strike a balance between entertainment and accuracy. This may involve consulting with experts in the relevant scientific fields, incorporating accurate terminology and concepts where possible, and providing educational materials or resources for viewers who wish to learn more.

Ultimately, while the inclusion of the multiverse in pop culture may present challenges in terms of scientific accuracy, it also offers valuable opportunities to inspire and educate audiences about the wonders of the universe and the cutting-edge research being conducted by physicists. By approaching the subject with care and respect for scientific principles, pop culture can continue to play a positive role in fostering scientific curiosity and understanding.

Philosophical Dilution: The portrayal of the multiverse in media can also dilute its profound philosophical implications, focusing more on the entertainment value than on the deeper existential questions that these theories raise.

- The portrayal of the multiverse in media often prioritizes entertainment value over the exploration of its profound philosophical implications, leading to a dilution of the deeper existential questions that these theories raise. While pop culture depictions of the multiverse can be engaging and thought-provoking, they may sometimes gloss over or simplify complex philosophical concepts in favor of spectacle, action, or humor.

One of the primary ways in which the philosophical implications of the multiverse are diluted in media is through the emphasis on visual spectacle and narrative excitement. Movies, television shows, and video games that feature the multiverse often focus on dazzling special effects, thrilling action sequences, and intricate plot twists, drawing viewers in with spectacle rather than engaging them in meaningful philosophical discourse.

Additionally, pop culture depictions of the multiverse may prioritize entertainment value over intellectual depth, sacrificing nuanced exploration of philosophical themes in favor of broad strokes and simplistic narratives. Complex concepts such as free will, determinism, identity, and existential angst may be reduced to clichés or stereotypes, robbing them of their richness and profundity.

Moreover, the fast-paced nature of media consumption can make it difficult for audiences to fully engage with the philosophical implications of the multiverse. In a world where attention spans are increasingly fragmented, pop culture depictions of the multiverse may be consumed as fleeting entertainment rather than opportunities for deep reflection and contemplation.

Furthermore, the commercial imperatives of the entertainment industry may incentivize producers and creators to prioritize marketability and accessibility over intellectual rigor and philosophical depth. As a result, pop culture depictions of the multiverse may be tailored to appeal to mass audiences, resulting in a watering-down of complex philosophical concepts in order to reach the widest possible audience.

Despite these challenges, there are opportunities for media to engage with the philosophical implications of the multiverse in more meaningful and substantive ways. By incorporating thoughtful dialogue, nuanced character development, and moral ambiguity into their narratives, creators can encourage audiences to grapple with the deeper existential questions raised by the multiverse.

Moreover, by providing supplemental materials, educational resources, and opportunities for discussion and debate, media creators can empower audiences to explore the philosophical implications of the multiverse on their own terms. Through interactive experiences, fan communities, and online forums, viewers can engage in thoughtful reflection and exchange ideas about the nature of reality, the meaning of existence, and the mysteries of the cosmos.

Ultimately, while the portrayal of the multiverse in media may sometimes dilute its profound philosophical implications, there are also opportunities for creators to engage with these themes in more meaningful and substantive ways. By approaching the subject with care, respect, and a willingness to explore difficult questions, media can play a valuable role in fostering intellectual curiosity and encouraging audiences to ponder the deeper mysteries of the universe.

Conclusion

The portrayal of the multiverse in popular culture is a double-edged sword. While it has undoubtedly popularized complex ideas and made them part of the global conversation, it often sacrifices depth and accuracy for broader appeal. As we continue to explore the multiverse in both fiction and reality, it remains crucial for educators, scientists, and creators to bridge the gap between entertainment and education, ensuring that these portrayals inspire curiosity and understanding rather than mere fascination.

In our next chapter, we will transition from the fictional representations of the multiverse to the real-world scientific endeavors and experimental setups aimed at testing the limits of our understanding of the universe. This will involve a deep dive into how current technologies and future experiments might one day confirm the existence of other universes or fundamentally alter our understanding of reality.

~ 6 ~

DREAMS AND PARALLEL LIVES

In this chapter, we venture into a more metaphysical exploration of the multiverse concept by examining the intriguing connection between dreams and the potential for parallel lives. We'll delve into the role of dreams as potential gateways to alternate realities, the scientific and psychic perspectives on dream experiences, and consider anecdotal and case studies that suggest deeper, possibly multidimensional aspects of dreaming.

The Nature of Dreams

Psychological Interpretations: Traditionally, dreams have been viewed through a psychological lens as manifestations of subconscious desires, fears, or unresolved conflicts. Sigmund Freud and Carl Jung were pioneers in theorizing that dreams reveal much about the inner workings of the mind, with Jung going further to suggest that they connect to a collective unconscious shared among all humans.

- Dreams have long captured the imagination of humanity, and throughout history, various cultures and scholars have interpreted them in diverse ways. One prominent approach to understanding dreams is through a psychological lens, which views dreams as manifestations of subconscious desires, fears, or unresolved conflicts. This perspective gained significant traction with the work of pioneering psychologists Sigmund Freud and Carl Jung.

Sigmund Freud, often regarded as the father of psychoanalysis, proposed that dreams are a window into the unconscious mind. In Freud's view, dreams serve as a means of wish fulfillment, allowing individuals to satisfy unconscious desires that are repressed or suppressed in waking life. According to Freud, dreams are filled with symbols and imagery that represent hidden wishes and conflicts, and the process of dream analysis involves uncovering these latent meanings through interpretation.

Freud's theory of dream interpretation posits that dreams contain both manifest content, which is the literal storyline of the dream, and latent content, which consists of the symbolic meanings underlying the manifest content. By analyzing the symbols, themes, and emotions present in dreams, Freud believed that individuals could gain insight into their innermost thoughts, feelings, and motivations.

Carl Jung, a Swiss psychiatrist and founder of analytical psychology, built upon Freud's work and introduced his own theories of dream interpretation. While Jung agreed with Freud that dreams are influenced by the unconscious mind, he expanded upon this idea by proposing the concept of the collective unconscious. According to Jung, the collective unconscious is a reservoir of shared symbolic imagery and archetypal motifs that are inherited and universal to all humans.

Jung believed that dreams tap into the collective unconscious, serving as a bridge between the personal unconscious and the collective unconscious. In Jungian dream analysis, the symbols and archetypes that appear in dreams are seen as reflections of deeper psychological truths and universal themes that resonate across cultures and time periods. By exploring these symbols and archetypes, individuals can gain insight into their own psyche and connect with broader aspects of the human experience.

Both Freudian and Jungian approaches to dream interpretation have had a profound influence on psychology, psychotherapy, and popular culture. While some aspects of Freud's theories have been criticized and revised over time, the idea that dreams offer valuable insights into the unconscious mind remains a central tenet of modern psychology. Similarly, Jung's concept of the collective unconscious continues to resonate with scholars and practitioners seeking to understand the deeper dimensions of human experience.

In conclusion, the psychological interpretation of dreams, as pioneered by Sigmund Freud and Carl Jung, offers a rich framework for understanding the complex symbolism and imagery present in dreams. By exploring the subconscious mind and its connections to the collective unconscious, individuals can gain profound insights into their innermost thoughts, feelings, and motivations, enriching their understanding of themselves and the human condition.

Neuroscientific Perspectives: Modern neuroscience offers insights into the brain mechanisms behind dreaming, particularly during the REM (Rapid Eye Movement) sleep stage. Studies show that dreams are associated with the processing and consolidation of memories, emotional regulation, and the random activation of neural pathways, which the brain attempts to weave into coherent narratives.

In recent years, advancements in neuroscience have provided valuable insights into the underlying brain mechanisms behind dreaming, particularly during the REM (Rapid Eye Movement) sleep stage. Through neuroimaging techniques such as functional magnetic resonance imaging (fMRI) and electroencephalography (EEG), researchers have been able to study the neural correlates of dreaming and shed light on the complex processes that occur in the brain during sleep.

One of the key findings from neuroscientific studies of dreaming is that it is closely linked to the processing and consolidation of memories. During REM sleep, which is when most vivid dreaming occurs, the brain is highly active, particularly in regions associated with memory formation and storage, such as the hippocampus and neocortex. Studies have shown that memories and experiences from waking life are reactivated and replayed during REM sleep, potentially facilitating the consolidation of new information and the integration of memories into existing knowledge networks.

Additionally, neuroscientific research has highlighted the role of dreaming in emotional regulation and processing. Studies have found that emotional experiences and memories are often incorporated into dreams, and the emotional content of dreams can influence waking mood and affect. During REM sleep, the amygdala, a brain region involved in emotion processing, shows increased activity, suggesting that dreaming may serve as a mechanism for processing and modulating emotional responses.

Furthermore, neuroscientific studies have revealed that dreaming involves the random activation of neural pathways and networks in the brain. During REM sleep, the brain experiences a state of increased neural activity and connectivity, characterized by bursts of electrical activity in the cortex. This heightened neural activity is thought to result in the spontaneous activation of memories, thoughts, and sensory experiences, which the brain attempts to weave into coherent narratives.

The phenomenon of random neural activation during dreaming is believed to give rise to the bizarre and surreal elements often observed in dreams. Dreams may consist of fragmented scenes, nonsensical events, and improbable scenarios, as the brain attempts to make sense of the disparate neural signals being generated during REM sleep.

Overall, neuroscientific perspectives on dreaming have greatly enhanced our understanding of the brain mechanisms underlying this complex phenomenon. By studying the neural correlates of dreaming, researchers have uncovered the role of REM sleep in memory consolidation, emotional processing, and the generation of dream content. While many questions remain unanswered, ongoing research in neuroscience continues to unravel the mysteries of dreaming and its significance for brain function and cognition.

Dreams as Multidimensional Experiences

Theories of Multidimensional Dreaming: Some metaphysicians and quantum physicists speculate that dreams might be more than just internal narratives. They propose that dreams could be experiences in parallel dimensions or alternate realities, where the dreamer's consciousness temporarily resides in another version of their life, accessing scenarios and outcomes different from their waking reality.

- The concept of multidimensional dreaming delves into the realm of metaphysics and quantum physics, offering speculative theories that suggest dreams may transcend the boundaries of our waking reality. Metaphysicians and some quantum physicists entertain the idea that dreams might be more than just internal narratives or random neural firings during sleep. Instead, they propose that dreams could be experiences in parallel dimensions or alternate realities, where the dreamer's consciousness temporarily resides in another version of their life, accessing scenarios and outcomes different from those experienced in their waking reality.

From a metaphysical perspective, proponents of multidimensional dreaming suggest that consciousness is not limited to the physical body or the constraints of space-time. Instead, they propose that consciousness exists across multiple dimensions or planes of existence, allowing individuals to access different realities or states of being during sleep. In this view, dreams serve as portals to alternate dimensions or parallel universes, where the dreamer's consciousness can explore different possibilities and experiences beyond the confines of their waking life.

Quantum physicists, drawing upon the principles of quantum mechanics, offer a more scientific interpretation of multidimensional dreaming. According to some interpretations of quantum theory, the universe consists of multiple dimensions or parallel realities that exist simultaneously, each with its own set of physical laws and possibilities. In this framework, consciousness is seen as a fundamental aspect of reality that can interact with and influence these parallel dimensions.

From a quantum perspective, dreams may be viewed as a form of quantum entanglement, where the dreamer's consciousness becomes entangled with the quantum states of alternate realities or parallel universes during sleep. This could potentially allow individuals to access information, experiences, and outcomes from these alternate realities, providing insights and perspectives that transcend the limitations of classical physics.

While theories of multidimensional dreaming remain speculative and largely outside the scope of mainstream scientific inquiry, they raise intriguing questions about the nature of consciousness, reality, and the relationship between mind and matter. Whether viewed through a metaphysical or quantum lens, the concept of multidimensional dreaming challenges conventional notions of dreams as purely internal phenomena, opening up new possibilities for exploring the mysteries of the human mind and its connection to the broader universe.

Quantum Mechanics and Dreaming: Building on the principles of quantum superposition and entanglement, some theories suggest that the human mind might interact with information from multiple universes during sleep. This hypothesis is bolstered by the non-locality principle of quantum physics, which could theoretically allow for the instantaneous connection between different states of consciousness spread across the multiverse.

- The intersection of quantum mechanics and dreaming presents a fascinating area of speculation that explores the possibility of the human mind interacting with information from multiple universes during sleep. Building on the principles of quantum superposition and entanglement, some theories propose that dreams may offer glimpses into parallel realities or alternate dimensions, where the dreamer's consciousness interacts with different states of existence.

  Quantum superposition, a fundamental principle of quantum mechanics, suggests that particles can exist in multiple states simultaneously until measured or observed. Similarly, proponents of quantum dreaming suggest that during sleep, the mind may exist in a state of superposition, simultaneously accessing information from multiple universes or alternate realities. In this view, dreams are not merely random neural firings or subconscious fantasies but rather a form of consciousness exploration across the multiverse.

Entanglement, another key concept in quantum mechanics, describes the phenomenon where particles become correlated with each other in such a way that the state of one particle is instantaneously influenced by the state of its entangled partner, regardless of the distance between them. Quantum dream theorists propose that the human mind may be entangled with the quantum states of alternate realities or parallel universes during sleep. This could allow for the exchange of information and experiences across different states of consciousness, transcending the limitations of classical physics.

The non-locality principle of quantum physics further supports the idea of interconnectedness between different states of consciousness spread across the multiverse. Non-locality suggests that particles can be instantaneously connected regardless of the distance between them, implying a form of interconnectedness that transcends conventional notions of space and time. In the context of quantum dreaming, this principle could theoretically allow for the instantaneous connection between different states of consciousness, enabling the exchange of information and experiences across parallel realities.

While theories of quantum dreaming remain speculative and largely theoretical, they raise provocative questions about the nature of consciousness, reality, and the relationship between mind and matter. By exploring the potential connections between quantum mechanics and dreaming, researchers and theorists are pushing the boundaries of scientific inquiry and challenging conventional understandings of the human experience. Whether viewed as a metaphorical interpretation or a literal exploration of the multiverse, the intersection of quantum mechanics and dreaming offers a rich and fertile ground for further exploration and speculation.

Case Studies and Anecdotal Evidence

Notable Experiences: Numerous individuals report dreams that are vivid, consistent, and detailed enough to suggest ongoing alternate lives. These accounts often describe environments, relationships, and events markedly different from the dreamers' actual lives but maintained with continuity and detail across different dreams.

- The phenomenon of individuals reporting dreams that seem to depict ongoing alternate lives is a fascinating aspect of human experience that has intrigued psychologists, philosophers, and researchers for centuries. These notable experiences often involve dreams that are vivid, consistent, and detailed enough to suggest the existence of parallel realities or alternate dimensions. In such dreams, individuals describe environments, relationships, and events that are markedly different from their actual lives but are maintained with continuity and detail across different dream episodes.

  One notable aspect of these experiences is the consistency and coherence of the dream narratives. Unlike typical dreams, which may be disjointed or fragmented, dreams depicting alternate lives often unfold in a linear and coherent manner, with clear storylines, characters, and settings. Individuals may find themselves inhabiting a different body, living in an unfamiliar place, or engaging in activities that are entirely foreign to their waking lives.

Moreover, these dreams often feature recurring themes, motifs, or locations that persist across multiple dream episodes. Individuals may report visiting the same alternate reality or encountering the same characters and situations in different dreams over time. This sense of continuity and familiarity lends credence to the idea that these dreams are not random creations of the subconscious mind but rather glimpses into parallel lives or alternate dimensions.

Another notable aspect of these experiences is the emotional intensity and psychological impact they can have on individuals. Dreams depicting alternate lives may evoke powerful emotions, ranging from joy and excitement to confusion and disorientation. Individuals may experience a profound sense of connection to the alternate reality depicted in their dreams, leading to feelings of longing, nostalgia, or even existential questioning about the nature of reality and identity.

In some cases, individuals may report experiencing a form of "dream déjà vu," where events, locations, or interactions from their dreams seem to overlap with experiences from their waking lives. This phenomenon can blur the boundary between dream and reality, leading individuals to question the distinction between the two and prompting them to explore the deeper meaning and significance of their dreams.

While the nature of these experiences remains open to interpretation, they raise intriguing questions about the nature of consciousness, the relationship between dreams and reality, and the possibility of parallel dimensions or alternate realities. Whether viewed as products of the subconscious mind, manifestations of psychological processes, or glimpses into alternate lives, dreams depicting parallel realities continue to captivate the imagination and inspire further exploration into the mysteries of human consciousness and existence.

Research and Documentation: Some researchers have begun systematically documenting such dreams, looking for patterns that might suggest a non-random, structured experience that could hint at parallel realities. While still highly speculative, these studies aim to apply a more rigorous scientific approach to what are often dismissed as merely vivid imaginations.

- The systematic documentation and study of dreams depicting alternate lives represent a growing area of research within the field of psychology and consciousness studies. While these experiences have traditionally been dismissed as products of vivid imagination or dismissed as anecdotal, some researchers have begun to take a more rigorous and scientific approach to understanding them. By systematically documenting these dreams and looking for patterns that might suggest a non-random, structured experience, researchers aim to explore the possibility that these dreams could offer insights into parallel realities or alternate dimensions.

  One approach taken by researchers involves collecting and analyzing dream reports from individuals who claim to have experienced dreams depicting alternate lives. These reports may be gathered through surveys, interviews, or online forums, allowing researchers to compile a database of accounts and identify common themes, motifs, or characteristics shared among them. By examining the content, frequency, and consistency of these dreams, researchers can begin to discern patterns that may shed light on the underlying nature of these experiences.

  In addition to qualitative analysis, some researchers are exploring quantitative methods for studying dreams depicting alternate lives. This may involve developing standardized measures or scales for assessing the intensity, coherence, and emotional impact of these dreams. By quantifying various aspects of the dream experience, researchers can compare and contrast different cases, identify correlations, and test hypotheses about the underlying mechanisms driving these phenomena.

Furthermore, advances in neuroimaging technology offer new opportunities for studying the neural correlates of dreams depicting alternate lives. By using techniques such as fMRI or EEG, researchers can investigate brain activity patterns during sleep and compare them to waking consciousness. This may provide insights into the neural mechanisms underlying the generation of these dreams and their relationship to other cognitive processes such as memory, imagination, and self-awareness.

While research on dreams depicting alternate lives is still in its infancy and remains highly speculative, it represents an exciting frontier in the study of consciousness and human experience. By applying a more rigorous scientific approach to these phenomena, researchers hope to gain a deeper understanding of the nature of dreams, the relationship between consciousness and reality, and the potential existence of parallel dimensions or alternate realities. While many questions remain unanswered, continued research in this area promises to shed light on one of the most intriguing mysteries of the human mind.

Psychic Perspectives on Dream Realities

Intuitive Dream Reading: In psychic practices, dreams are often interpreted as messages from the subconscious or as glimpses into alternate dimensions of existence. Psychics and spiritual practitioners sometimes use dreams as tools for guidance, healing, and insight into one's parallel lives or paths not taken.

• Intuitive dream reading is a practice rooted in various psychic and spiritual traditions, where dreams are interpreted as messages from the subconscious mind or glimpses into alternate dimensions of existence. In these practices, psychics, mediums, and spiritual practitioners use their intuition and spiritual insight to interpret the symbolism, themes, and imagery present in dreams, offering guidance, healing, and insight into the dreamer's life and potential paths not taken.

One of the key principles underlying intuitive dream reading is the belief that dreams contain valuable information and insights that can help individuals navigate their lives and understand their innermost thoughts, feelings, and desires. In this view, dreams are seen as a form of communication between the conscious and subconscious mind, offering clues and guidance that may not be readily apparent in waking life.

Psychics and spiritual practitioners often employ various techniques and methods to interpret dreams intuitively. These may include meditation, visualization, tarot card readings, astrology, or other divinatory practices that help them tap into their intuition and access deeper levels of awareness. By connecting with the dreamer's energy and the symbolism present in the dream, intuitive interpreters can offer insights and guidance that resonate with the dreamer's unique circumstances and life path.

In addition to providing insight into the dreamer's current life circumstances, intuitive dream reading may also offer glimpses into parallel lives or alternate dimensions of existence. Some practitioners believe that dreams can transcend the boundaries of time and space, allowing individuals to access information and experiences from past lives, future potentials, or parallel realities. By exploring these alternate dimensions through dreams, individuals may gain a broader perspective on their own life journey and the interconnectedness of all existence.

Furthermore, intuitive dream reading can be used as a tool for healing and personal growth. By exploring the symbolism and themes present in dreams, individuals can uncover hidden patterns, unresolved emotions, and unconscious desires that may be influencing their waking life. Through the process of dream interpretation, individuals can gain clarity, insight, and empowerment to make positive changes in their lives and align with their highest purpose and potential.

While intuitive dream reading is often regarded as a subjective and non-scientific practice, many individuals find value and meaning in the insights and guidance offered by psychics and spiritual practitioners. Whether viewed as a form of divination, psychological exploration, or spiritual communion, intuitive dream reading provides a unique perspective on the mysteries of the human psyche and the interconnectedness of consciousness with the greater universe.

Therapeutic Applications: Beyond traditional dream analysis, some therapeutic practices incorporate the idea of multidimensional dreaming into their approaches, helping individuals explore these 'other lives' as a means to gain insight into their current challenges or life choices.

- Therapeutic applications of multidimensional dreaming go beyond traditional dream analysis to incorporate the idea that dreams may offer glimpses into parallel realities or alternate dimensions of existence. In these approaches, therapists and practitioners use multidimensional dreaming as a tool for exploration and self-discovery, helping individuals gain insight into their current challenges, life choices, and personal growth.

One therapeutic approach that incorporates multidimensional dreaming is known as dream exploration therapy. In this approach, therapists work with clients to explore the symbolism, themes, and imagery present in their dreams, with the understanding that dreams may contain messages or insights from parallel realities or alternate dimensions. By analyzing and interpreting these dreams within the context of the client's life circumstances and psychological processes, therapists can help clients gain a deeper understanding of themselves and their experiences.

Dream exploration therapy may involve techniques such as guided visualization, journaling, dream recall exercises, and dream interpretation sessions. Through these practices, clients can uncover hidden patterns, unconscious beliefs, and unresolved issues that may be influencing their waking life. By exploring the symbolism and themes present in their dreams, clients can gain clarity, insight, and empowerment to make positive changes in their lives and overcome obstacles to personal growth.

Another therapeutic approach that incorporates multidimensional dreaming is known as transpersonal therapy. Transpersonal therapy explores the spiritual dimensions of human experience, including the idea that consciousness may extend beyond the boundaries of the individual self to encompass broader aspects of the universe. In this approach, dreams are seen as a window into the soul and a pathway to higher states of consciousness.

In transpersonal therapy, therapists work with clients to explore the spiritual significance of their dreams, including any experiences of multidimensional awareness or connections to alternate realities. By integrating the insights gained from these experiences into the therapeutic process, clients can deepen their understanding of themselves, their relationships, and their place in the world. Transpersonal therapy may involve practices such as meditation, mindfulness, and energy work to help clients access deeper levels of consciousness and explore the mysteries of the human psyche.

Overall, therapeutic applications of multidimensional dreaming offer a holistic approach to healing and personal growth, integrating insights from psychology, spirituality, and metaphysics. By exploring the symbolic language of dreams and the potential connections to parallel realities, individuals can gain a deeper understanding of themselves and their experiences, leading to greater self-awareness, inner peace, and fulfillment in life.

Conclusion

The exploration of dreams as potential connections to parallel universes invites us to reconsider the boundaries between reality and imagination, between science and metaphysics. While definitive proof remains elusive, the continued study of dreams and their possible implications for understanding the multiverse reflects our enduring quest to comprehend the full spectrum of human experience and consciousness.

As we close this chapter, we remain mindful of the profound mystery surrounding dreams and their role in our lives. Whether they truly connect us to parallel realities or are simply the mind's way of processing our waking life, dreams undeniably hold a key to deeper self-understanding and, perhaps, to the infinite possibilities of the cosmos. In the next chapter, we will delve into the phenomenon of déjà vu and its potential connections to the multiverse, exploring yet another aspect of our lives where the ordinary can sometimes touch the extraordinary.

~ 7 ~

DÉJÀ VU AND THE MULTIVERSE

In this chapter, we delve into the enigmatic phenomenon of déjà vu, exploring its potential connections to parallel universes and the multiverse theory. This phenomenon, commonly experienced as an eerie sense of familiarity with a new situation, has puzzled scientists, psychologists, and metaphysicians alike. We will examine scientific explanations, psychic interpretations, and how déjà vu might serve as an intersection between our perceived reality and alternate timelines.

Understanding Déjà Vu

Scientific Explanations: In the realm of neuroscience, déjà vu is often attributed to a glitch in the brain's memory system, similar to a short circuit that causes the brain to mistake the present for the past. Researchers suggest that it occurs when there's a temporal misalignment in the sensory input and retrieval of a memory, leading to the uncanny feeling that an event or experience has happened before.

- In the realm of neuroscience, déjà vu, the eerie sensation of having experienced a present situation before, is

82

often attributed to various theories rooted in the workings of the brain's memory system. While déjà vu has intrigued researchers and puzzled individuals for centuries, scientific explanations seek to demystify this phenomenon by proposing mechanisms within the brain that could give rise to the sensation.

One prevalent theory posited by neuroscientists suggests that déjà vu occurs due to a temporary glitch or malfunction in the brain's memory processing system. According to this explanation, the brain's memory network involves complex interactions between different regions, including the hippocampus, responsible for encoding new memories, and the neocortex, involved in higher-order cognitive processes. During déjà vu, researchers propose that there is a brief disruption or short circuit in this memory system, leading to a momentary confusion between the present sensory input and stored memories.

More specifically, researchers suggest that déjà vu may arise when there is a temporal misalignment between the sensory input from the current experience and the retrieval of a similar memory from the past. In other words, the brain may incorrectly retrieve a memory from the past that closely resembles the present experience, leading to the uncanny feeling that the current situation has been previously encountered. This temporal misalignment could be due to various factors, such as rapid processing of information, overlapping neural pathways, or interference from similar memories.

Neuroscientific studies have provided some support for this theory by identifying brain regions and neural processes that may be involved in generating the experience of déjà vu. For example, neuroimaging studies have shown increased activity in areas of the medial temporal lobe, including the hippocampus and surrounding regions, during déjà vu experiences. This suggests that the brain's memory network may play a central role in generating the sensation of déjà vu by erroneously retrieving and processing memories from the past.

Additionally, researchers have proposed alternative explanations for déjà vu, including dysfunction in the brain's familiarity detection mechanisms or errors in the perception of time. Some studies have suggested that déjà vu may be more common in individuals with certain neurological conditions, such as epilepsy or migraine, further highlighting the potential role of brain dysfunction in generating this phenomenon.

While scientific explanations of déjà vu provide valuable insights into the underlying mechanisms within the brain, the phenomenon remains complex and multifaceted, with many aspects still not fully understood. Continued research into the neural basis of déjà vu may yield further insights into the workings of memory and cognition, shedding light on one of the mind's most intriguing mysteries.

Psychological Perspectives: Psychologists have explored the role of subconscious memory and processing in triggering déjà vu experiences. It's posited that subliminal recognition of patterns or cues in the environment may lead to this phenomenon, without the conscious mind realizing why the situation feels familiar.

- Psychological perspectives on déjà vu delve into the workings of the subconscious mind and its role in triggering this intriguing phenomenon. Psychologists have long been fascinated by déjà vu and have proposed various theories to explain its occurrence, focusing on subconscious memory and processing as key factors.

One prominent theory within psychology suggests that déjà vu arises from the subliminal recognition of patterns or cues in the environment. According to this perspective, individuals may encounter stimuli or situations that bear similarities to past experiences, even if those similarities are not consciously recognized. These subtle cues may include familiar sights, sounds, smells, or other sensory inputs that trigger memories stored in the subconscious mind.

Psychologists propose that when individuals experience déjà vu, their subconscious mind perceives these familiar cues in the environment and retrieves associated memories from long-term storage. However, because the recognition occurs at a subconscious level, individuals may not consciously recall the original experiences or understand why the current situation feels familiar. Instead, they may simply experience the uncanny sensation that they have been in the same place or situation before, without being able to pinpoint why.

This theory of subconscious recognition aligns with research on memory and cognition, which suggests that the brain is constantly processing vast amounts of information at both conscious and subconscious levels. Studies have shown that individuals can subliminally perceive and process information even when they are not consciously aware of it, leading to subtle influences on behavior, perception, and memory.

Furthermore, psychologists have explored the role of memory retrieval processes in generating déjà vu experiences. According to some theories, déjà vu may occur when there is a mismatch or discrepancy between the speed of memory retrieval and the processing of incoming sensory information. In other words, the brain may retrieve a memory from the past that closely resembles the present experience, but the conscious mind lags behind in recognizing the source of familiarity.

Psychological perspectives on déjà vu highlight the complex interplay between perception, memory, and consciousness in shaping our subjective experience of reality. While the phenomenon of déjà vu remains enigmatic and elusive, research from psychological perspectives offers valuable insights into the workings of the human mind and the intricate mechanisms underlying our perception of the world. By exploring the role of subconscious memory and processing, psychologists continue to unravel the mysteries of déjà vu and its significance for understanding cognition, memory, and consciousness.

Déjà Vu and the Multiverse Theory

Alternate Realities: From a metaphysical perspective, some theorists propose that déjà vu might be moments when our consciousness briefly aligns with the experience of an alternate version of ourselves in a parallel universe. This theory suggests that our multiversal selves share certain experiences across different dimensions, and déjà vu is a fleeting perception of these shared moments.

- The concept of alternate realities, also known as parallel universes or alternate dimensions, has long fascinated metaphysical thinkers, philosophers, and theorists. From this perspective, some theorists propose that déjà vu experiences may offer glimpses into moments when our consciousness briefly aligns with the experiences of an alternate version of ourselves in a parallel universe. This intriguing theory suggests that our multiversal selves share certain experiences across different dimensions, and déjà vu is a fleeting perception of these shared moments.

According to this metaphysical perspective, the multiverse is envisioned as a vast and interconnected web of parallel realities, each existing simultaneously but independently of one another. In this framework, every individual has countless counterparts or "doppelgängers" inhabiting different dimensions, each living out their own unique lives and making different choices. Despite the separation between these parallel selves, there may be moments when their experiences converge or overlap, leading to a sense of familiarity or recognition in the form of déjà vu.

Proponents of this theory suggest that déjà vu occurs when our consciousness momentarily resonates with the experiences of our counterparts in alternate realities. During these moments, the barriers between dimensions may become temporarily blurred, allowing for the transmission of information or energy between parallel universes. As a result, individuals may perceive echoes or fragments of experiences that belong to their alternate selves, leading to the uncanny sensation that they have lived through the same moment before.

From this perspective, déjà vu is seen as a form of cross-dimensional communication or resonance, where the boundaries between different versions of ourselves become temporarily porous. This theory suggests that our multiversal selves may share common threads of experience, emotion, or consciousness that transcend the limitations of individual identity and existence. Déjà vu experiences, therefore, may serve as reminders of our interconnectedness with the broader fabric of the multiverse and the vastness of our own potentiality across different dimensions.

While this metaphysical interpretation of déjà vu remains speculative and beyond the realm of empirical science, it offers a thought-provoking perspective on the nature of consciousness, reality, and existence. By exploring the concept of alternate realities, theorists seek to expand our understanding of the universe and the possibilities that lie beyond the boundaries of conventional perception. Whether viewed as a fleeting glimpse into parallel dimensions or a trick of the mind, déjà vu continues to captivate the imagination and inspire contemplation about the mysteries of the human experience.

Quantum Mechanics and Memory: Building on the principles of quantum entanglement and superposition, there's a speculative idea that human memory might occasionally access information from multiple quantum states. In essence, this could mean that memories from parallel lives seep into our current consciousness, presenting as déjà vu.

- The intersection of quantum mechanics and memory introduces a speculative idea that human memory might occasionally access information from multiple quantum states, potentially leading to phenomena such as déjà vu. This concept builds upon the principles of quantum entanglement and superposition, which describe how particles can exist in multiple states simultaneously and how the states of entangled particles are correlated regardless of distance.

  In this speculative framework, human memory is conceptualized as a complex and dynamic system that may interact with quantum phenomena at the microscopic level. According to quantum theory, particles can exist in a state of superposition, meaning they can be in multiple states at once, until they are observed or measured. Similarly, some theorists propose that memories stored in the brain may exist in a state of quantum superposition, allowing them to access information from multiple quantum states simultaneously.

One key concept in this theory is the idea of quantum entanglement, which suggests that particles can become correlated in such a way that the state of one particle is instantaneously influenced by the state of its entangled partner, regardless of the distance between them. Applied to human memory, this idea suggests that memories from different quantum states may become entangled or interconnected, leading to the phenomenon of déjà vu.

According to this speculative theory, déjà vu occurs when the brain retrieves memories that are entangled with memories from parallel quantum states. In essence, memories from parallel lives or alternate realities may seep into our current consciousness, leading to the uncanny sensation that we have experienced a particular moment before. Because these memories are entangled with our own, they may feel familiar or recognizable, even though they belong to a different version of ourselves in a parallel universe.

While this theory remains highly speculative and lacks empirical evidence, it offers a provocative framework for understanding the mysteries of memory and consciousness from a quantum perspective. By exploring the potential connections between quantum mechanics and memory, theorists seek to unravel the intricate relationship between the mind and the fabric of reality. Whether viewed as a metaphorical interpretation or a literal exploration of quantum phenomena, the idea of quantum memory provides fertile ground for further inquiry and speculation into the nature of human experience.

Case Studies and Research

Documented Instances: While anecdotal, numerous accounts of déjà vu come from individuals who describe these experiences as too specific and detailed to be dismissed as mere memory glitches. These instances often involve knowing intricate details about places they have never visited or predicting what happens next in a completely unfamiliar setting.

- Documented instances of déjà vu provide intriguing insights into the complexity and enigmatic nature of this phenomenon. While anecdotal, these accounts come from individuals who describe their experiences as too specific and detailed to be dismissed as mere memory glitches or coincidences. Instead, these instances often involve knowing intricate details about places they have never visited or predicting what happens next in a completely unfamiliar setting, leading to a sense of bewilderment and fascination.

  One common theme among documented instances of déjà vu is the vividness and clarity of the experience. Individuals often report feeling as though they are reliving a moment from their past, with sensory details and emotions that are exceptionally vivid and intense. These experiences may include not only visual images but also sounds, smells, and tactile sensations, adding to the sense of realism and authenticity.

  Moreover, documented instances of déjà vu frequently involve scenarios where individuals possess knowledge or information that they could not have acquired through ordinary means. For example, individuals may find themselves in a new place or situation and suddenly realize that they know intricate details about the surroundings, such as the layout of a building or the placement of objects, despite never having been there before. In some cases, individuals may even be able to predict future events or outcomes with remarkable accuracy, leading to a sense of prescience or clairvoyance.

One striking aspect of documented instances of déjà vu is the emotional impact they can have on individuals. Many people describe feeling a profound sense of disorientation, confusion, or déjà vécu ("already lived") during these experiences, as though they are caught in a loop of time or space. Some individuals may also experience feelings of nostalgia, déjà entendu ("already heard"), or déjà senti ("already felt"), adding layers of complexity to the déjà vu phenomenon.

Furthermore, documented instances of déjà vu challenge conventional explanations based solely on memory glitches or neurological anomalies. While these factors may play a role in some cases of déjà vu, they cannot fully account for the specificity, detail, and predictive accuracy of many reported experiences. Instead, these instances suggest that déjà vu may involve complex interactions between memory, perception, consciousness, and possibly even phenomena beyond our current understanding.

Overall, documented instances of déjà vu highlight the need for further research and exploration into this intriguing phenomenon. By documenting and analyzing these experiences in detail, researchers can gain valuable insights into the underlying mechanisms and potential implications of déjà vu for our understanding of memory, consciousness, and the nature of reality itself.

Experimental Approaches: Some experimental psychologists have attempted to recreate déjà vu in laboratory settings to understand its mechanisms better. These studies often use virtual reality or guided imagery to induce feelings of familiarity, though replicating the spontaneous nature of déjà vu remains challenging.

• Experimental approaches to studying déjà vu involve efforts by psychologists to recreate this phenomenon in laboratory settings in order to better understand its mechanisms and underlying causes. While challenging due to the spontaneous and unpredictable nature of déjà vu experiences, researchers have employed various techniques, including virtual reality and guided imagery, to induce feelings of familiarity and simulate aspects of déjà vu.

One common approach used in experimental studies of déjà vu involves the use of virtual reality technology. Researchers create immersive virtual environments that mimic real-world settings, such as city streets, buildings, or rooms. Participants are then exposed to these virtual environments and asked to navigate or explore them while their physiological responses, such as heart rate and skin conductance, are monitored. By manipulating certain elements within the virtual environment, such as the layout of buildings or the placement of objects, researchers can attempt to induce feelings of familiarity or recognition similar to those experienced during déjà vu.

Another experimental approach involves the use of guided imagery or suggestion to induce déjà vu-like experiences in participants. In these studies, participants are guided through a series of mental exercises or visualization techniques designed to evoke feelings of familiarity or recognition. For example, participants may be asked to imagine themselves in a familiar location or to recall a specific memory from their past. Researchers then measure participants' subjective experiences and physiological responses to assess the effectiveness of the induction technique.

Despite these experimental approaches, replicating the spontaneous nature of déjà vu remains a significant challenge. Déjà vu experiences often occur unexpectedly and are difficult to predict or control in laboratory settings. Additionally, individual differences in susceptibility to déjà vu and the subjective nature of the phenomenon make it challenging to standardize experimental protocols and measure outcomes consistently across participants.

Despite these challenges, experimental studies of déjà vu have provided valuable insights into the cognitive and neurological mechanisms underlying this phenomenon. For example, research using functional neuroimaging techniques has identified brain regions, such as the hippocampus and temporal lobes, that are involved in the generation of déjà vu experiences. Other studies have explored the role of memory processes, attentional mechanisms, and perceptual cues in triggering déjà vu.

Overall, while experimental approaches to studying déjà vu face numerous challenges, they offer valuable opportunities to investigate the mechanisms and causes of this intriguing phenomenon. By combining innovative methodologies with rigorous scientific inquiry, researchers continue to advance our understanding of déjà vu and its implications for memory, consciousness, and perception.

Psychic Perspectives on Déjà Vu

Insights from Psychic Practitioners: In the psychic community, déjà vu is sometimes viewed as proof of a person's ability to perceive beyond the linear progression of time. Psychics and spiritualists may interpret déjà vu as messages or signs from the universe, urging an individual to pay attention to their current life choices or circumstances.

- Insights from psychic practitioners offer a unique perspective on déjà vu, often viewing it as evidence of a person's ability to perceive beyond the linear progression of time and as a potential message or sign from the universe. Within the psychic community, déjà vu is interpreted through the lens of spirituality, intuition, and metaphysical beliefs, rather than purely as a psychological or neurological phenomenon.

Psychics and spiritualists often believe that déjà vu experiences are not random occurrences but are instead meaningful messages or signs from the universe. From this perspective, déjà vu may serve as a form of guidance or intuition, urging individuals to pay attention to their current life choices, circumstances, or paths. Psychics may interpret the recurrence of déjà vu as a signal that an individual is on the right path or that they are being guided toward a particular course of action.

Furthermore, psychic practitioners may view déjà vu as evidence of the interconnectedness of all existence and the existence of a larger cosmic plan or order. From this perspective, déjà vu experiences are seen as moments when individuals briefly tap into higher levels of consciousness or awareness, gaining insights or glimpses into the broader fabric of reality. Psychics may interpret these experiences as indications of synchronicity or alignment with universal forces, encouraging individuals to trust in the inherent wisdom of the universe.

In addition to providing guidance and insight, psychic practitioners may offer techniques or practices to help individuals enhance their intuitive abilities and interpret their déjà vu experiences more effectively. These techniques may include meditation, visualization, energy work, or other spiritual practices designed to expand consciousness and deepen connection with higher realms of existence.

Overall, insights from psychic practitioners offer a holistic and spiritually oriented perspective on déjà vu, viewing it as a meaningful and potentially transformative experience rather than simply a curious anomaly of the mind. By embracing the idea that déjà vu may hold deeper significance beyond its surface appearance, individuals can explore new avenues of self-discovery, growth, and spiritual awakening. Whether viewed as a message from the universe, a glimpse into parallel dimensions, or a manifestation of higher consciousness, déjà vu continues to intrigue and inspire contemplation about the mysteries of the human experience and the nature of reality itself.

Therapeutic Implications: Beyond its mysterious nature, déjà vu is sometimes used in therapeutic settings as a tool to explore unconscious memories or unresolved past traumas. The feeling of familiarity is examined for deeper emotional connections that might be influencing a person's present behavior or emotional state.

- Therapeutic implications of déjà vu extend beyond its enigmatic nature, as it can serve as a valuable tool in exploring unconscious memories or unresolved past traumas within therapeutic settings. Therapists may leverage the feeling of familiarity associated with déjà vu to delve into deeper emotional connections that could be influencing a person's present behavior, emotions, or mental well-being.

  In therapy, déjà vu experiences are explored within the broader context of a person's psychological history and current life circumstances. Therapists may encourage individuals to reflect on the specific details and emotions surrounding their déjà vu episodes, examining any patterns or themes that emerge. By exploring the origins and significance of these experiences, therapists can help clients gain insights into underlying emotional issues or unresolved conflicts that may be contributing to their present difficulties.

One therapeutic approach that utilizes déjà vu as a tool for exploration is psychoanalytic therapy. In this approach, therapists help clients uncover unconscious memories, thoughts, and emotions that may be influencing their behavior and relationships. Déjà vu experiences are seen as potential gateways to the unconscious mind, offering clues to unresolved conflicts or traumas that may be buried beneath conscious awareness. Through techniques such as free association, dream analysis, and interpretation of symbolic imagery, therapists guide clients in uncovering the deeper meanings behind their déjà vu episodes.

Another therapeutic implication of déjà vu is its potential use in trauma therapy. Traumatic experiences from the past can often remain hidden in the unconscious mind, manifesting as feelings of unease, anxiety, or déjà vu in the present. Therapists may use déjà vu experiences as entry points to explore these hidden traumas and help clients process and integrate them into their conscious awareness. By acknowledging and addressing unresolved traumas, individuals can experience healing and emotional growth, leading to greater resilience and well-being.

Overall, déjà vu can serve as a powerful therapeutic tool for exploring unconscious memories, unresolved traumas, and deeper emotional connections within the therapeutic process. By integrating déjà vu experiences into therapy sessions, therapists can help clients gain deeper self-awareness, resolve past conflicts, and achieve greater emotional healing and integration. Through this process, individuals can move toward greater self-understanding, empowerment, and fulfillment in their lives.

Conclusion

While the scientific community continues to explore the neurological and cognitive bases of déjà vu, the possibility that it might connect us to parallel dimensions offers an intriguing intersection of science and metaphysics. Whether viewed through the lens of psychology, neuroscience, or spiritual and psychic theory, déjà vu challenges our understanding of time, memory, and personal identity.

As we move forward in our exploration of the multiverse, the concept of déjà vu serves as a compelling example of how ordinary human experiences might hint at extraordinary truths about the nature of reality. In the next chapter, we will explore further how parallel lives might be experienced and understood through the stories and testimonies of those who believe they have encountered their other selves.

~ 8 ~

PARALLEL LIVES AMONG US

In this chapter, we delve into the fascinating concept of parallel lives, exploring the possibility that alongside our known existence, alternate versions of ourselves live out different scenarios in separate dimensions. This chapter not only investigates the theoretical and scientific underpinnings of such phenomena but also considers personal anecdotes and psychological implications, offering a comprehensive examination of how parallel lives might intersect with and influence our own.

Theoretical Foundations

Quantum Foundations: Building on the principles discussed in earlier chapters, we revisit the notion that every decision might spawn a new universe. Quantum mechanics suggests that for every possible outcome to an event, the universe splits, creating parallel worlds where each possibility becomes reality.

- In quantum foundations, the notion that every decision might spawn a new universe is a fascinating concept that stems from the principles of quantum mechanics. This idea, often associated with the many-worlds interpretation (MWI) of quantum mechanics, proposes that the universe splits into multiple parallel worlds or branches with each possible outcome of a quantum event becoming reality.

At the heart of this concept is the principle of superposition, which states that quantum particles can exist in multiple states simultaneously until they are observed or measured. According to quantum mechanics, when a measurement is made on a quantum system, it collapses into one of its possible states, leading to the observer witnessing a single outcome. However, proponents of the many-worlds interpretation suggest that instead of collapsing into a single outcome, the universe splits into multiple branches, each corresponding to a different possible outcome of the measurement.

For example, consider the classic thought experiment of Schrödinger's cat, where a cat inside a sealed box is both alive and dead until the box is opened and the cat's state is observed. In the many-worlds interpretation, the universe splits into two branches: one where the cat is alive and another where the cat is dead. Each of these branches represents a separate reality or parallel universe, coexisting alongside our own.

The implications of the many-worlds interpretation extend beyond the realm of quantum mechanics and into the macroscopic world of everyday experience. According to this interpretation, every quantum event, from the decay of a radioactive atom to the outcome of a coin toss, spawns a multitude of parallel universes, each corresponding to a different possible outcome. As a result, the number of parallel worlds proliferates exponentially with each quantum event, leading to an infinitely branching multiverse.

From a philosophical perspective, the many-worlds interpretation raises profound questions about the nature of reality, free will, and the observer's role in shaping the universe. If every possible outcome of a quantum event occurs in a separate parallel universe, then every decision we make, no matter how trivial, leads to the creation of new universes with different versions of ourselves and our reality.

While the many-worlds interpretation remains a topic of debate among physicists and philosophers, it offers a captivating glimpse into the potential richness and complexity of the quantum universe. By revisiting the notion that every decision might spawn a new universe, we are reminded of the inherent strangeness and beauty of quantum mechanics and its profound implications for our understanding of the cosmos.

Multiverse Scenarios: Theories from physics propose various types of multiverses—from the Many-Worlds Interpretation to the Bubble Universes theory of cosmic inflation. Each model provides a framework for understanding how parallel lives could exist, separated by cosmic or quantum boundaries.

- Multiverse scenarios, as theorized in physics, encompass a diverse range of models that propose the existence of multiple universes beyond our own. These theories offer intriguing frameworks for understanding how parallel lives and realities could exist, separated by cosmic or quantum boundaries. Among the prominent multiverse models are the Many-Worlds Interpretation (MWI) and the Bubble Universes theory of cosmic inflation.

1. Many-Worlds Interpretation (MWI): The Many-Worlds Interpretation is a hypothesis within quantum mechanics that suggests the existence of an infinite number of parallel universes. According to MWI, every quantum event results in the splitting of the universe

into multiple branches, each corresponding to a different outcome of the event. For example, in the famous Schrödinger's cat thought experiment, where the cat is both alive and dead until observed, MWI proposes that the universe bifurcates into one branch where the cat is alive and another where it is dead. In this framework, every possible outcome of a quantum event is realized in a separate universe, leading to a vast and branching multiverse.

2.   Bubble Universes Theory: The Bubble Universes theory stems from the concept of cosmic inflation, which proposes that the early universe underwent a rapid exponential expansion shortly after the Big Bang. According to this theory, regions of space can continue to undergo inflationary expansion, leading to the formation of "bubble" universes within a larger multiverse. Each bubble universe is separated from others by vast expanses of space and may have different physical constants or laws of physics. These bubbles are thought to arise from quantum fluctuations during the inflationary period, giving rise to an ensemble of universes with varying properties.

Other multiverse scenarios proposed in physics include the concept of a "landscape" of string theory, where the configuration of extra dimensions gives rise to a vast array of possible universes with different physical laws and properties. Additionally, theories such as the "parallel worlds" model in quantum cosmology and the "brane-world" scenario in string theory posit the existence of parallel universes or higher-dimensional realms beyond our own.

Overall, multiverse scenarios in physics offer captivating insights into the potential complexity and diversity of the cosmos. By exploring different models of the multiverse, physicists aim to deepen our understanding of the fundamental nature of reality and our place within the larger cosmic tapestry. While these theories remain speculative and subject to ongoing research and debate, they provide fertile ground for exploring the profound implications of parallel lives and universes on the fabric of existence.

Anecdotal Evidence

Personal Stories: Numerous individuals report experiences that suggest interactions with parallel worlds. These stories often involve strong feelings of déjà vu, unexplained knowledge about unfamiliar places, or vivid dreams that depict alternative life paths.

- Personal stories recounting interactions with parallel worlds offer fascinating glimpses into the potential intersections between different realities. While anecdotal, these accounts often describe experiences that evoke strong feelings of déjà vu, unexplained knowledge about unfamiliar places, or vivid dreams that depict alternative life paths. These stories can be both intriguing and thought-provoking, prompting individuals to consider the possibility of parallel lives and universes beyond our own.

1. Déjà Vu Experiences: Many personal stories of interactions with parallel worlds involve episodes of déjà vu, where individuals feel a sudden and inexplicable sense of familiarity with a particular situation or location. These experiences can be intense and vivid, with individuals experiencing a strong sensation of reliving a moment from their past, despite never having been in that specific situation before. For some, déjà vu serves

as a glimpse into a parallel reality where the same event or scenario has occurred previously, leading to feelings of wonder and curiosity about the nature of reality.

2. Unexplained Knowledge: Another common theme in personal stories is the acquisition of unexplained knowledge about unfamiliar places or events. Individuals may suddenly possess detailed information about a location they have never visited or have insights into future events that seem to come out of nowhere. These experiences can be disconcerting yet intriguing, leading individuals to question the source of this newfound knowledge and consider the possibility of tapping into parallel dimensions or alternate realities.

3. Vivid Dreams and Alternative Life Paths: Some personal stories involve vivid dreams that depict alternative life paths or scenarios vastly different from one's waking reality. In these dreams, individuals may find themselves living out different roles, making different choices, or experiencing events that diverge from their actual life trajectory. While dreams are often dismissed as products of the subconscious mind, some individuals interpret these vivid dream experiences as glimpses into parallel worlds where alternative versions of themselves exist.

Overall, personal stories of interactions with parallel worlds offer compelling narratives that challenge conventional notions of reality and consciousness. While skeptics may dismiss these accounts as mere flights of imagination or coincidence, others find resonance in the idea that our experiences may extend beyond the confines of our known universe. Whether viewed as metaphors for personal transformation or glimpses into the multiverse, these stories invite contemplation and exploration of the mysteries of existence.

Historical and Cultural Perspectives: Across cultures, there are myths and legends that hint at the existence of parallel worlds. From the Norse tales of Yggdrasil linking different realms to the spiritual beliefs in shadow selves found in various indigenous cultures, these narratives can be interpreted as early understandings of parallel lives.

- Historical and cultural perspectives provide rich insights into the concept of parallel worlds, as evidenced by myths, legends, and spiritual beliefs from diverse cultures around the world. These narratives offer intriguing glimpses into early understandings of parallel lives and alternative realities, reflecting humanity's enduring fascination with the mysteries of existence.

1. Norse Mythology and the Nine Realms: In Norse mythology, the cosmology is depicted as a complex interconnected web of nine realms, each inhabited by different beings and governed by distinct cosmic forces. Central to this cosmology is Yggdrasil, the World Tree, which serves as a cosmic axis linking the nine realms together. These realms include Asgard, home of the gods; Midgard, the realm of humans; and Jotunheim, the land of the giants. The existence of multiple realms in Norse mythology suggests a belief in parallel worlds or dimensions, each with its own inhabitants and characteristics.

2. Indigenous Spiritual Beliefs: Many indigenous cultures around the world hold spiritual beliefs that hint at the existence of parallel worlds or alternate realities. For example, in various indigenous traditions, there are beliefs in shadow selves or spirit doubles that

exist alongside an individual's physical body. These shadow selves are thought to inhabit parallel worlds or spiritual realms and may influence a person's actions, thoughts, and experiences. Additionally, indigenous creation myths often describe the emergence of multiple worlds or realms from a primordial chaos, reflecting a worldview that embraces the interconnectedness of all existence.

3. Eastern Philosophical Traditions: In Eastern philosophical traditions such as Hinduism and Buddhism, there are concepts that resonate with the idea of parallel worlds or dimensions. In Hindu cosmology, for example, the universe is depicted as a vast multiverse consisting of countless realms or lokas, each inhabited by different gods, goddesses, and celestial beings. These realms exist within the broader framework of cyclical cosmic cycles of creation, preservation, and dissolution, known as Yugas. Similarly, in Buddhist cosmology, there are descriptions of multiple planes of existence, including realms of gods, humans, animals, and spirits, each characterized by different levels of consciousness and karma.

4. Shamanic Practices: In shamanic traditions found in various cultures worldwide, shamans are believed to have the ability to journey between different realms or worlds in altered states of consciousness. Through rituals, drumming, chanting, and other techniques, shamans seek to traverse the boundaries between the physical world and the spirit realms, accessing hidden knowledge, healing energies, and guidance from spiritual beings. These practices suggest a belief in the existence of parallel worlds or dimensions that can be explored and interacted with through spiritual means.

Overall, historical and cultural perspectives on parallel worlds offer diverse and nuanced understandings of the concept, reflecting humanity's enduring quest to comprehend the mysteries of existence and our place within the cosmos. By exploring these myths, legends, and spiritual beliefs, we gain valuable insights into the ways different cultures have grappled with questions of reality, consciousness, and the nature of existence throughout history.

Scientific Investigations

Experimental Approaches: While direct evidence of parallel lives is elusive, scientists are exploring ways to test the reality of parallel universes. This includes examining cosmic radiation patterns that might indicate collisions of our universe with others, and using quantum computers to simulate conditions that could reveal characteristics of multiversal interactions.

- Experimental approaches aimed at investigating the reality of parallel universes represent a cutting-edge frontier in scientific inquiry. While direct evidence of parallel lives remains elusive, scientists are actively exploring various methodologies and techniques to test the existence of parallel universes and uncover potential signs of multiversal interactions. Some of these experimental approaches include:

1. Cosmic Radiation Patterns: One approach involves analyzing cosmic radiation patterns in an effort to detect evidence of collisions between our universe and others. According to certain cosmological theories, such collisions could leave distinctive imprints on the cosmic microwave background (CMB) radiation, which

permeates the universe. Scientists study subtle variations and anomalies in the CMB to search for signatures that might indicate interactions with parallel universes. By analyzing data from large-scale surveys and experiments such as the Planck satellite and the Atacama Cosmology Telescope, researchers aim to identify potential anomalies that could provide clues about the existence of parallel universes.

2. Quantum Computing Simulations: Another approach involves using quantum computers to simulate conditions that could reveal characteristics of multiversal interactions. Quantum computers leverage the principles of quantum mechanics to perform computations at speeds and scales that surpass those of classical computers. Researchers are exploring the potential of quantum computing to model complex quantum systems, including those that may be indicative of parallel universes. By simulating scenarios involving entanglement, superposition, and quantum decoherence, scientists hope to gain insights into the behavior of quantum systems across multiple universes and uncover potential signatures of multiversal phenomena.

3. Particle Collider Experiments: Particle collider experiments conducted at facilities such as the Large Hadron Collider (LHC) at CERN also offer opportunities to probe for evidence of parallel universes. These experiments involve accelerating subatomic particles to high energies and colliding them together at nearly the speed of light. Scientists analyze the resulting particle interactions and decay patterns to search for deviations from standard model predictions that could indicate the presence of extra dimensions, hidden particles, or other phenomena associated with parallel universes. While no direct evidence has been found thus far,

ongoing experiments continue to push the boundaries of particle physics and cosmology.

4. Multidisciplinary Approaches: In addition to these experimental techniques, researchers are exploring multidisciplinary approaches that combine insights from cosmology, particle physics, quantum mechanics, and other fields to address questions related to parallel universes. Collaborative efforts among scientists from diverse disciplines aim to develop novel theoretical frameworks, observational methods, and technological innovations for probing the nature of reality at the deepest levels.

Overall, experimental approaches to investigating parallel universes represent a dynamic and evolving area of scientific exploration. While significant challenges remain, ongoing research efforts hold the potential to shed light on one of the most profound and enigmatic questions in cosmology: the existence and nature of parallel lives and universes beyond our own.

Psychological Analysis: Psychologists study the impact of beliefs in parallel lives on human behavior. This research considers how such beliefs might affect decision-making, anxiety about alternate choices, and the overall mental health of individuals.

• Psychological analysis of beliefs in parallel lives provides valuable insights into how these beliefs influence human behavior, decision-making processes, and mental health outcomes. Psychologists explore the cognitive, emotional, and behavioral ramifications of holding such beliefs, considering their implications for individual well-being and functioning within society. Here are some key aspects of psychological analysis in this area:

1. Decision-Making Processes: Beliefs in parallel lives can profoundly influence decision-making processes by introducing the concept of alternative outcomes and possibilities. Individuals who hold such beliefs may experience heightened awareness of the potential consequences of their choices, leading to increased deliberation and introspection during decision-making. This heightened sensitivity to alternative outcomes may manifest as decisional conflict or anxiety, as individuals grapple with the uncertainty inherent in choosing one path over another. Psychologists examine how beliefs in parallel lives shape individuals' decision-making strategies, risk perceptions, and preferences for certain outcomes over others.

2. Anxiety and Stress: Beliefs in parallel lives can also contribute to feelings of anxiety and stress, particularly when individuals ruminate on the myriad possibilities and scenarios that could have unfolded differently in their lives. The concept of parallel lives may exacerbate existential concerns about the nature of reality, free will, and the inevitability of certain life events. Individuals may experience anxiety about the potential ramifications of their choices in one reality compared to another, leading to feelings of regret, uncertainty, or existential angst. Psychologists explore how beliefs in parallel lives intersect with anxiety disorders, stress responses, and coping mechanisms for managing uncertainty and existential concerns.

3. Coping Mechanisms and Resilience: Despite the potential challenges posed by beliefs in parallel lives, individuals may also develop adaptive coping mechanisms

and resilience strategies to navigate existential uncertainty and anxiety. Psychologists investigate how individuals reconcile conflicting beliefs about determinism and free will, finding meaning and purpose in their lives, and cultivating resilience in the face of existential dilemmas. Beliefs in parallel lives may serve as a source of comfort or empowerment for some individuals, offering a sense of agency and possibility amidst life's uncertainties. Psychologists explore the role of cognitive reframing, perspective-taking, and acceptance in fostering resilience and psychological well-being in individuals who hold such beliefs.

4. Cultural and Individual Differences: Psychological analysis considers the role of cultural and individual differences in shaping beliefs in parallel lives and their psychological impact. Cultural factors such as religious beliefs, philosophical traditions, and societal norms may influence the prevalence and interpretation of beliefs in parallel lives within different populations. Additionally, individual differences in personality traits, cognitive styles, and emotional regulation strategies may modulate the psychological effects of holding such beliefs. Psychologists examine how cultural context and individual variability shape the expression and consequences of beliefs in parallel lives across diverse populations.

Overall, psychological analysis of beliefs in parallel lives provides valuable insights into the complex interplay between cognition, emotion, and behavior in individuals' perceptions of reality and their quest for meaning and purpose in life. By understanding the psychological mechanisms underlying these beliefs, psychologists can offer support, guidance, and interventions to promote psychological well-being and resilience in individuals who grapple with existential questions and uncertainties about the nature of existence.

Psychic and Metaphysical Insights

Psychic Experiences: Many psychics and mediums claim to have the ability to perceive connections to parallel lives. Through readings and spiritual sessions, they offer insights that purportedly link individuals to their alternate existences, providing guidance or closure on unresolved issues.

- Psychic experiences, as reported by psychics and mediums, involve purported connections to parallel lives or alternate realities, offering insights, guidance, and closure on unresolved issues. These experiences are often described as intuitive or extrasensory perceptions that transcend conventional sensory channels, allowing psychics to access information from realms beyond the physical world. While controversial and subject to skepticism, psychic experiences have been reported across cultures and historical periods, playing a role in spiritual and metaphysical traditions worldwide. Here are key aspects of psychic experiences related to parallel lives:

1. Intuitive Insights: Psychics and mediums claim to receive intuitive insights and impressions about individuals' parallel lives through various means, such as clairvoyance (visions), clairaudience (auditory messages), or clairsentience (emotional or physical sensations). During readings or spiritual sessions, psychics may describe vivid images, sensations, or messages that they attribute to connections with parallel realities. These insights are often personalized and may provide

individuals with a sense of validation, comfort, or clarity regarding their experiences and life circumstances.

2. Past Life Regression: Some psychics specialize in past life regression, a technique aimed at accessing memories and experiences from past incarnations or parallel lives. Through hypnosis or guided meditation, individuals are led to explore memories and sensations that purportedly originate from previous lifetimes. Psychics guide clients through these experiences, helping them uncover insights, patterns, or unresolved issues that may be influencing their current life situation. Past life regression sessions are often framed as opportunities for healing, self-discovery, and spiritual growth.

3. Mediumship and Communication with Spirits: Mediums claim to communicate with spirits or entities that exist in parallel realms or dimensions beyond the physical world. During mediumship sessions, mediums purportedly receive messages, impressions, or symbols from deceased loved ones, spirit guides, or other spiritual beings. These communications may provide individuals with reassurance, guidance, or closure regarding unresolved issues or relationships. Mediums often describe their ability to bridge the gap between the physical and spiritual realms, facilitating connections between individuals and their parallel lives or departed loved ones.

4. Guidance and Closure: Psychic experiences related to parallel lives often focus on providing guidance, validation, or closure for individuals seeking answers to existential questions or unresolved dilemmas. Psychics may offer insights into past-life connections, karmic patterns, or soul contracts that influence individuals' current life circumstances. By accessing information from parallel realities, psychics aim to empower

individuals to make informed decisions, heal emotional wounds, and find peace or resolution in their lives.

5.   Skepticism and Criticism: Despite the popularity of psychic experiences, they remain controversial and subject to skepticism from scientists, skeptics, and critics. Skeptics argue that psychic phenomena are often based on subjective interpretations, cold reading techniques, or cognitive biases rather than genuine connections to parallel lives or spiritual realms. Critics raise concerns about the potential for exploitation, manipulation, and deception in the psychic industry, urging caution when seeking guidance from psychics or mediums.

Overall, psychic experiences related to parallel lives offer a unique perspective on the nature of consciousness, reality, and the interconnectedness of existence. While they may provide comfort or validation for some individuals, skepticism and critical inquiry remain essential for evaluating the validity and ethical implications of psychic claims.

Therapeutic Uses: Some therapeutic practices incorporate the idea of parallel lives to help clients deal with regrets or explore untapped potential. Techniques such as guided meditations or hypnotherapy are used to access what some believe to be memories from parallel existences.

- Therapeutic uses of the concept of parallel lives involve integrating this idea into therapeutic practices to help clients explore unresolved issues, overcome regrets, and tap into untapped potential. Therapists employ various techniques, such as guided meditations, hypnotherapy, and narrative therapy, to facilitate exploration of what some believe to be memories or insights from parallel existences. Here are key aspects of therapeutic approaches incorporating parallel lives:

1. Guided Meditations: Therapists may guide clients through meditation practices aimed at accessing memories, insights, or experiences from parallel lives. During guided meditations, clients are encouraged to relax, focus their attention inward, and visualize themselves exploring alternate realities or lifetimes. Therapists may prompt clients to imagine specific scenarios, encounters, or decisions from their parallel lives, encouraging reflection, insight, and emotional processing. Guided meditations can serve as a tool for self-exploration, healing, and personal growth, allowing clients to gain new perspectives on their current life circumstances and choices.

2. Hypnotherapy and Past Life Regression: Hypnotherapy and past life regression techniques are employed to facilitate exploration of parallel lives and past incarnations. Under the guidance of a trained therapist, clients enter a hypnotic trance state where they are believed to access subconscious memories and experiences from parallel existences. Therapists may use suggestive cues, prompts, or regression scripts to guide clients through past-life memories, emotions, and sensations. Past life regression sessions are often framed as opportunities for healing unresolved traumas, gaining insights into recurring patterns or relationships, and exploring untapped talents or passions from previous lifetimes.

3. Narrative Therapy and Alternative Perspectives: In narrative therapy, clients are encouraged to explore alternative perspectives, stories, or identities that may exist within themselves. Therapists work collaboratively with clients to co-create narratives that incor-

porate elements of parallel lives, alternate realities, or imagined selves. Through storytelling, metaphor, and creative expression, clients can externalize their experiences, beliefs, and aspirations, gaining distance from entrenched patterns or self-limiting narratives. Narrative therapy provides a safe and empowering space for clients to reimagine their identities, relationships, and life trajectories, drawing inspiration from the concept of parallel lives as a source of resilience and possibility.

4. Integration and Meaning-Making: Therapeutic approaches incorporating parallel lives emphasize the importance of integrating insights and experiences from alternate realities into clients' current life narratives. Therapists help clients make meaning of their parallel life experiences, identifying themes, lessons, and strengths that can inform their present-day choices and actions. By integrating insights from parallel lives, clients can develop a deeper understanding of themselves, cultivate self-compassion, and embrace new possibilities for growth and transformation.

Overall, therapeutic uses of the concept of parallel lives offer clients a unique framework for exploring unresolved issues, reframing regrets, and envisioning alternative paths forward. While these approaches may be speculative in nature, they can provide valuable opportunities for self-reflection, healing, and personal empowerment within the therapeutic context. Therapists guide clients on a journey of self-discovery and transformation, drawing upon the rich potential of parallel lives as a source of insight, resilience, and growth.

Implications for Personal Growth

Understanding Self and Potential: Exploring the concept of parallel lives can lead to deeper self-awareness and a broader understanding of one's potential. It encourages individuals to reflect on the multitude of choices they make and the possible impacts these choices have across different dimensions.

- Exploring the concept of parallel lives can indeed lead to profound insights into the self and the realization of one's potential. By contemplating the idea that alternate versions of oneself exist in parallel realities, individuals are prompted to engage in deep introspection and reflection on the choices, decisions, and possibilities that shape their lives. Here are several ways in which exploring parallel lives can foster greater understanding of the self and one's potential:

1. Self-Reflection and Choice Awareness: Contemplating parallel lives encourages individuals to reflect on the myriad choices they make in their daily lives and the potential consequences of these choices across different dimensions. By examining the paths not taken and the outcomes that could have unfolded differently, individuals gain a heightened awareness of their decision-making processes and the factors influencing their choices. This self-reflection fosters greater mindfulness and intentionality in decision-making, empowering individuals to align their actions with their values, goals, and aspirations.

2. Embracing Complexity and Multifaceted Identities: Exploring parallel lives invites individuals to embrace the complexity of their identities and the multiplicity of selves that exist within them. Rather than viewing the self as a singular, static entity, individuals come to

recognize themselves as dynamic, multifaceted beings with the capacity for growth, adaptation, and transformation. This expanded understanding of the self encourages individuals to embrace their diverse interests, talents, and potentials, recognizing that different versions of themselves may thrive in different contexts or realities.

3. Cultivating Empathy and Perspective-Taking: Contemplating parallel lives fosters empathy and perspective-taking by encouraging individuals to consider the experiences, perspectives, and challenges faced by their alternate selves. By empathizing with the struggles and triumphs of their parallel counterparts, individuals gain a deeper appreciation for the complexities of human existence and the interconnectedness of all beings. This heightened empathy enhances interpersonal relationships, fosters compassion, and promotes a sense of solidarity with others across different realities.

4. Navigating Uncertainty and Embracing Possibility: Exploring parallel lives equips individuals with the resilience and adaptability needed to navigate life's uncertainties and challenges. By recognizing that reality is inherently probabilistic and that multiple futures are possible, individuals develop a sense of openness to new experiences, opportunities, and possibilities. This mindset shift enables individuals to embrace change, take calculated risks, and pursue novel pathways with confidence and curiosity.

5. Actualizing Potential and Personal Growth: Ultimately, contemplating parallel lives empowers individuals to actualize their potential and pursue personal growth with purpose and intention. By envisioning the diverse trajectories their lives could take, individuals are inspired to set ambitious goals, overcome obstacles,

and strive for excellence in all aspects of their lives. This proactive approach to personal development fosters resilience, self-efficacy, and a sense of fulfillment as individuals harness their innate talents and passions to create meaningful lives aligned with their deepest aspirations.

In summary, exploring the concept of parallel lives offers a transformative journey of self-discovery and empowerment, enabling individuals to gain deeper insights into their identities, values, and potentials. By embracing complexity, cultivating empathy, and navigating uncertainty with resilience, individuals unlock new possibilities for growth, fulfillment, and self-actualization across multiple dimensions of existence.

Ethical and Moral Considerations: The belief in parallel lives raises intriguing ethical questions. If multiple versions of us exist, making different choices, how does this affect our understanding of responsibility and morality?

- The belief in parallel lives indeed raises intriguing ethical and moral considerations that challenge traditional notions of responsibility, accountability, and morality. Exploring the implications of parallel lives prompts individuals to grapple with complex philosophical questions about free will, determinism, and the nature of ethical decision-making. Here are several key ethical and moral considerations associated with the belief in parallel lives:

1. Responsibility and Accountability: The concept of parallel lives challenges conventional understandings

of responsibility and accountability by suggesting that different versions of oneself may make divergent choices and face distinct consequences in alternate realities. Individuals may question the extent to which they are morally responsible for the actions and outcomes of their parallel counterparts. This raises ethical dilemmas about culpability, agency, and the distribution of moral responsibility across multiple selves.

2. Moral Relativism and Pluralism: Belief in parallel lives introduces the possibility of moral relativism and pluralism, wherein ethical principles and values may vary across different dimensions or realities. Individuals may confront the idea that what is considered morally right or wrong in one reality may differ from another, leading to relativistic perspectives on ethics and morality. This challenges individuals to navigate ethical dilemmas with humility, empathy, and an openness to diverse moral frameworks.

3. Impact on Relationships and Interactions: The belief in parallel lives can influence individuals' attitudes and behaviors in their relationships and interactions with others. Individuals may question the significance of their choices and actions in shaping the lives of their parallel counterparts, leading to reflections on empathy, compassion, and the interconnectedness of all beings. This awareness of parallel lives may foster greater empathy, forgiveness, and understanding in interpersonal relationships as individuals recognize the complexity of human experience.

4. Ethical Decision-Making Across Realities: Exploring parallel lives prompts individuals to consider the ethical implications of their choices and actions across different dimensions. Individuals may confront ethical dilemmas about whether to prioritize their own well-

being, pursue altruistic goals, or consider the potential impacts of their decisions on their parallel counterparts. This raises questions about moral decision-making in a multiverse context and the ethical principles that guide individuals' behavior across multiple realities.

5. Existential Ethics and Meaning-Making: Belief in parallel lives can provoke existential reflections on the nature of ethics, meaning, and purpose in life. Individuals may grapple with questions about the ultimate significance of their choices and experiences in the vast expanse of the multiverse. This existential perspective invites individuals to contemplate their values, aspirations, and contributions to the greater good, seeking to find meaning and purpose amidst the complexity of parallel realities.

In summary, the belief in parallel lives engenders profound ethical and moral considerations that challenge individuals to reexamine their understanding of responsibility, accountability, and morality. By confronting the implications of multiple selves making different choices across alternate realities, individuals are prompted to engage in ethical reflection, moral deliberation, and existential inquiry, fostering a deeper understanding of the complexities of human existence and the ethical dimensions of life in a multiverse.

Conclusion

The notion of parallel lives challenges our conventional understanding of reality and identity. As we consider the intersections between our lives and potential alternate existences, we are compelled to reconsider the boundaries of our personal and collective narratives. This chapter invites readers to explore the profound implications of living in a multiverse—not just as a theoretical curiosity but as a transformative perspective on life's possibilities.

In the next chapter, we will turn our attention to the scientific skepticism and support for the multiverse theory, exploring the debates and discussions that shape current scientific thought on this expansive topic.

$$\sim 9 \sim$$

SCIENTIFIC SKEPTICISM AND SUPPORT

In this chapter, we delve into the contentious debate surrounding the multiverse theory, examining both the scientific skepticism and support that this concept has garnered within the academic community. By exploring the arguments on both sides, we aim to provide a balanced view that acknowledges the complexities and limitations of current scientific understanding while also highlighting the theoretical advancements that support the possibility of multiple universes.

The Roots of Skepticism

Lack of Empirical Evidence: The primary critique of the multiverse theory stems from its lack of direct empirical evidence. Critics argue that without observable and measurable data, the theory veers more into the realm of philosophical speculation rather than empirical science.

- The lack of empirical evidence is indeed a primary critique leveled against the multiverse theory, posing significant challenges to its scientific credibility and testability. Critics contend that without tangible, observable evidence to support the existence of parallel universes, the theory remains speculative and speculative, straying from the foundational principles of empirical science. Here are several key points regarding the critique of the multiverse theory's lack of empirical evidence:

1. Verification and Falsifiability: A cornerstone of empirical science is the principle of verification and falsifiability, wherein scientific hypotheses must be testable through empirical observation and experimentation. Critics argue that the multiverse theory, by positing the existence of parallel universes beyond the reach of direct observation, lacks the necessary empirical basis for verification or falsifiability. Without the ability to test the theory through observable data, skeptics contend that it falls outside the domain of empirical science.

2. Methodological Limitations: Critics point to the methodological limitations inherent in attempting to empirically detect or measure parallel universes. The vast distances, cosmic scales, and hypothetical properties of parallel realities present formidable challenges to observational and experimental inquiry. Current scientific instruments and technologies may be insufficient to detect phenomena associated with parallel universes, further complicating efforts to gather empirical evidence in support of the multiverse theory.

3. Occam's Razor and Parsimony: Another criticism of the multiverse theory's lack of empirical evidence revolves around the principle of Occam's razor, which suggests that simpler explanations should be favored over more complex ones when accounting for observed phenomena. Critics argue that invoking the existence of an infinite number of parallel universes to explain

observed phenomena may violate the principle of parsimony, as it introduces unnecessary complexity without empirical justification.

4. Alternative Explanations: Skeptics of the multiverse theory propose alternative explanations for observed phenomena that do not rely on the existence of parallel universes. These alternative hypotheses may invoke principles from established scientific theories, such as quantum mechanics, cosmology, or string theory, to account for observed phenomena within the framework of a single universe. Critics argue that these alternative explanations offer more parsimonious and empirically grounded accounts of reality than the multiverse theory.

5. Philosophical Speculation: Finally, critics contend that the multiverse theory's lack of empirical evidence relegates it to the realm of philosophical speculation rather than empirical science. While the theory may inspire thought experiments, metaphysical inquiries, and speculative conjectures, its inability to be tested or confirmed through empirical observation undermines its status as a scientifically viable hypothesis.

In summary, the lack of empirical evidence poses a significant challenge to the multiverse theory's scientific credibility and testability. Critics argue that without tangible observational or experimental data to support its claims, the theory remains speculative and speculative, falling short of the rigorous empirical standards required for acceptance within the scientific community.

The Problem of Falsifiability: Karl Popper's principle of falsifiability states that for a theory to be considered scientific, it must be testable and potentially falsifiable. Skeptics point out that the multiverse theory, as currently presented, often lacks clear predictions that could be tested to prove the theory false, making it problematic from a scientific methodology standpoint.

- The problem of falsifiability, as applied to the multiverse theory, raises significant concerns regarding its scientific status and testability within the framework of Karl Popper's principle of falsifiability. According to Popper, a scientific theory must be formulated in a way that allows for empirical testing and the possibility of being falsified through observation or experimentation. However, skeptics argue that the multiverse theory, in its current formulations, often lacks clear predictions or criteria that could be tested to potentially refute or falsify the theory. Here are several key points regarding the problem of falsifiability as it relates to the multiverse theory:

1. Lack of Observable Predictions: One of the primary criticisms leveled against the multiverse theory is its purported lack of observable predictions that could be tested through empirical observation or experimentation. Critics contend that many formulations of the multiverse theory propose the existence of parallel universes or dimensions that are beyond the reach of direct observation, rendering them inherently untestable and unfalsifiable. Without specific predictions or criteria that could be empirically verified or refuted, skeptics argue that the multiverse theory fails to meet the criteria of falsifiability set forth by Popper.

2. Unobservable Entities and Phenomena: Another aspect of the problem of falsifiability in the multiverse theory concerns the reliance on unobservable entities

and phenomena to support its claims. Many formulations of the multiverse theory invoke concepts such as parallel universes, alternate dimensions, or cosmic inflation beyond the observable universe, which are not directly accessible to empirical observation or measurement. Critics argue that theories relying on unobservable entities or phenomena risk straying into the realm of metaphysics rather than empirical science, as they cannot be subjected to empirical testing or verification.

3. Challenges in Experimental Design: Even if specific predictions could be derived from the multiverse theory, skeptics argue that practical challenges may arise in designing experiments or observational studies to test these predictions. The vast distances, cosmic scales, and hypothetical properties associated with parallel universes or alternate dimensions present formidable obstacles to empirical investigation. Critics contend that the sheer complexity and speculative nature of the multiverse theory make it difficult, if not impossible, to design experiments that could definitively confirm or refute its claims.

4. Theoretical Unfalsifiability: Some critics of the multiverse theory assert that certain formulations of the theory may be inherently unfalsifiable due to their speculative and metaphysical nature. For example, theories positing an infinite number of parallel universes with varying physical laws or properties may lack clear criteria for falsification, as virtually any observed phenomenon could potentially be accommodated within the framework of the theory. Critics argue that such theories risk becoming immune to empirical testing or refutation, undermining their status as scientifically viable hypotheses.

In summary, the problem of falsifiability poses a significant challenge to the scientific credibility and testability of the multiverse theory. Critics argue that without specific predictions that could be empirically tested or falsified, the theory risks straying into the realm of metaphysics rather than empirical science. Addressing the problem of falsifiability may require refining the formulations of the multiverse theory to generate testable predictions or criteria that could be subjected to empirical scrutiny within the framework of Popper's principle of falsifiability.

Occam's Razor: This philosophical principle suggests that when presented with competing hypotheses, the one with the fewest assumptions should be selected. Critics argue that the multiverse theory introduces unnecessary complexity into explanations of the universe, contrary to the simplicity favored by Occam's Razor.

- Occam's Razor, also known as the principle of parsimony or the law of economy, is a fundamental philosophical principle that guides scientific inquiry and hypothesis formation. It suggests that among competing hypotheses or explanations, the one that makes the fewest assumptions or postulates is to be preferred. In other words, simpler explanations are generally favored over more complex ones, all else being equal. However, critics argue that the multiverse theory, with its invocation of parallel universes and additional dimensions, violates Occam's Razor by introducing unnecessary complexity into explanations of the universe. Here's how critics frame their argument:

1. Unnecessary Multiplicity: One of the key criticisms of the multiverse theory from the perspective of Occam's

Razor is its introduction of unnecessary multiplicity. Rather than positing a single, unified explanation for observed phenomena, the multiverse theory suggests the existence of an infinite number of parallel universes or dimensions, each with its own set of physical laws and properties. Critics argue that this proliferation of parallel realities adds unnecessary complexity to explanations of the universe, as it requires postulating the existence of entities and phenomena beyond what is strictly necessary to account for observed data.

2. Complexity Without Justification: Critics further contend that the multiverse theory introduces complexity without sufficient justification. While proponents of the multiverse theory invoke it to explain certain observed phenomena, such as the fine-tuning of physical constants or the anthropic principle, critics argue that simpler, more elegant explanations may suffice. They maintain that invoking an infinite number of parallel universes as an explanatory mechanism is unwarranted, as it introduces speculative entities and phenomena that cannot be empirically verified or falsified.

3. Lack of Empirical Evidence: Another criticism of the multiverse theory in the context of Occam's Razor is its reliance on speculative entities and phenomena without empirical evidence to support them. Critics argue that Occam's Razor favors explanations grounded in empirical observation and testable predictions, whereas the multiverse theory often lacks direct observational or experimental evidence. Without empirical verification, skeptics contend that the multiverse theory's invocation of parallel universes violates Occam's Razor by introducing unnecessary complexity based on speculative assumptions rather than empirical data.

4.   Philosophical and Methodological Considerations: Critics also raise philosophical and methodological objections to the multiverse theory's departure from Occam's Razor. They argue that Occam's Razor is not merely a heuristic or guideline but a foundational principle of scientific inquiry that underscores the importance of simplicity, elegance, and explanatory power in scientific theories. By invoking parallel universes or dimensions to explain observed phenomena, critics assert that the multiverse theory deviates from the principles of parsimony and economy favored by Occam's Razor, undermining its status as a scientifically viable hypothesis.

In summary, critics of the multiverse theory argue that it violates Occam's Razor by introducing unnecessary complexity into explanations of the universe. They contend that simpler, more elegant explanations should be preferred unless compelling empirical evidence warrants the introduction of additional entities or phenomena. By challenging the multiverse theory on grounds of Occam's Razor, skeptics highlight the importance of simplicity, parsimony, and empirical verification in scientific theorizing.

Support from Theoretical Physics

Cosmic Inflation Theory: Supporters of the multiverse draw on aspects of cosmic inflation, an extension of the Big Bang theory, which suggests that 'bubble universes' could form during the rapid expansion of space. This aspect of inflationary theory offers a framework where different universes could have different physical laws, supporting the multiverse concept.

- Cosmic inflation theory, an extension of the Big Bang model, proposes that the early universe underwent a period of exponential expansion shortly after the Big Bang. This rapid expansion, driven by a hypothetical field called the inflaton, is believed to have occurred within a fraction of a second after the Big Bang, stretching the fabric of space-time and smoothing out irregularities in the universe. Supporters of the multiverse theory draw on aspects of cosmic inflation to provide a framework for the existence of parallel universes or "bubble universes." Here's how cosmic inflation theory supports the concept of the multiverse:

1. Rapid Expansion and Bubble Universes: During the inflationary epoch, the universe experienced an extraordinary burst of expansion, increasing its size by an enormous factor in a very short period. This rapid expansion led to the formation of "bubble universes" within a larger, inflating space-time. In this scenario, each bubble universe is a self-contained region with its own set of physical laws and properties, effectively operating as a separate universe.

2. Quantum Fluctuations and Bubble Formation: According to cosmic inflation theory, quantum fluctuations in the early universe played a crucial role in the formation of bubble universes. Quantum fluctuations are random fluctuations in energy at the quantum level, arising from the inherent uncertainty principle of quantum mechanics. These fluctuations can lead to the spontaneous creation of regions with slightly different energy densities, resulting in the formation of bubble universes within the expanding space-time.

3. Eternal Inflation and Multiverse: In some models of cosmic inflation, the inflationary process never completely stops in certain regions of space-time, leading

to the concept of eternal inflation. In this scenario, inflation continues indefinitely in some regions, while inflation ceases and gives rise to observable universes like our own in other regions. The ongoing process of eternal inflation generates a vast ensemble of bubble universes, each with its own unique set of physical laws and properties. This proliferation of bubble universes forms the basis for the multiverse concept.

4.   Variation in Physical Laws: One of the key implications of the multiverse scenario arising from cosmic inflation is the possibility of variation in physical laws and constants across different bubble universes. Since each bubble universe forms independently within the larger inflating space-time, they can exhibit different properties, such as different values for fundamental constants or different types of particles. This variation in physical laws provides a natural framework for the existence of parallel universes with diverse characteristics.

In summary, cosmic inflation theory offers a compelling framework for the existence of parallel universes or bubble universes within the multiverse concept. By positing a period of rapid expansion and the spontaneous formation of bubble universes during the early universe, cosmic inflation theory provides a plausible mechanism for the generation of diverse universes with different physical laws and properties, supporting the idea of a vast multiverse.

String Theory Landscapes: String theory posits that different configurations of string vibrations could result in multiple stable forms of space-time, potentially leading to different physical constants and laws. The landscape of string theory naturally lends itself to a multiverse interpretation, providing a mathematical foundation that supports the existence of diverse universes.

- String theory is a theoretical framework in physics that attempts to reconcile general relativity (describing gravity) with quantum mechanics (describing the other three fundamental forces of nature) by modeling elementary particles as one-dimensional "strings" rather than point particles. One of the intriguing features of string theory is its prediction of multiple possible configurations of string vibrations, each corresponding to different stable forms of space-time. These configurations, often referred to as "string theory landscapes," provide a mathematical framework that naturally lends itself to a multiverse interpretation. Here's how string theory landscapes support the concept of the multiverse:

1. Vibrational Modes and Space-Time Configurations: In string theory, the vibrational modes of strings determine the properties of elementary particles and the structure of space-time. Different vibrational modes correspond to different particle types and interactions, as well as different geometrical configurations of space-time. These configurations can include various dimensions, shapes, and topologies, allowing for a rich diversity of possible space-time structures.
2. Stability and Vacuum States: Within string theory, certain configurations of string vibrations correspond to stable vacuum states or "ground states" of the theory. These vacuum states represent possible forms of space-time with stable physical properties, including the values of fundamental constants and the laws governing particle interactions. The landscape of string theory encompasses the full range of possible vacuum

states, each characterized by its own set of physical laws and properties.

3. Multiverse Interpretation: The existence of multiple stable vacuum states in string theory naturally leads to a multiverse interpretation. Each vacuum state corresponds to a potential universe with its own distinct features, such as different particle masses, force strengths, and spacetime geometries. The vast array of possible vacuum states in the string theory landscape suggests the existence of a diverse ensemble of universes, collectively forming a multiverse.

4. Anthropic Principle: The concept of the anthropic principle is often invoked in the context of string theory landscapes to explain the apparent fine-tuning of physical constants and the emergence of life-permitting conditions in our universe. According to the anthropic principle, the observed values of fundamental constants and physical laws in our universe are not arbitrary but are instead constrained by the requirement for the existence of observers like ourselves. In the multiverse scenario of string theory landscapes, the anthropic principle suggests that our universe's properties may be a result of the particular vacuum state in which it resides, among the multitude of possible vacuum states in the landscape.

In summary, string theory landscapes provide a mathematical framework that supports the existence of a multiverse by predicting multiple stable forms of space-time with different physical laws and properties. The diversity of vacuum states in the string theory landscape suggests the existence of a vast ensemble of universes, each with its own unique characteristics, contributing to the concept of a multiverse in theoretical physics.

Quantum Mechanics: The many-worlds interpretation of quantum mechanics, which posits that all possible outcomes of quantum measurements are physically realized in some "world" or universe, provides a quantum mechanical basis for the multiverse. This interpretation eliminates the need for wave function collapse, offering a more deterministic approach to quantum physics that supports the multiverse theory.

- Quantum mechanics, the branch of physics that describes the behavior of particles at the atomic and subatomic scales, has revolutionized our understanding of the fundamental nature of reality. One of the most intriguing interpretations of quantum mechanics is the many-worlds interpretation (MWI), which proposes that every possible outcome of a quantum measurement is realized in a separate branch of reality or "world." Here's how the many-worlds interpretation provides a quantum mechanical basis for the multiverse concept:

1. Wave Function and Superposition: In quantum mechanics, particles such as electrons and photons are described by wave functions, mathematical entities that represent the probabilities of finding the particles in different states. According to the principle of superposition, a particle can exist in multiple states simultaneously until it is measured, at which point the wave function "collapses" to a single state corresponding to the measurement outcome. However, in the many-worlds interpretation, the wave function never collapses; instead, all possible outcomes of a measurement are realized in separate branches of reality.

2. Branching Universes: In the many-worlds interpretation, each possible outcome of a quantum measurement results in the creation of a new branch of reality or universe. For example, if a particle can be in two different states simultaneously, such as spin up and spin down, the universe splits into two branches, one where the particle is observed to be spin up and another where it is observed to be spin down. This branching process occurs continuously for every quantum interaction, leading to an exponentially growing number of parallel universes.

3. Elimination of Wave Function Collapse: One of the key features of the many-worlds interpretation is the elimination of the need for wave function collapse. In traditional interpretations of quantum mechanics, the collapse of the wave function upon measurement is seen as a fundamental and mysterious aspect of quantum reality. However, in the many-worlds interpretation, the wave function evolves deterministically according to the Schrödinger equation, with no collapse occurring. Instead, the branching of universes preserves the superposition of states and allows all possible outcomes to be realized.

4. Deterministic Approach: The many-worlds interpretation offers a more deterministic approach to quantum mechanics, where the evolution of the wave function follows predictable rules without the need for probabilistic wave function collapse. Each branch of the multiverse evolves independently according to the laws of quantum mechanics, leading to a coherent and deterministic picture of reality. This deterministic framework aligns well with the concept of the multiverse, where every possible outcome of quantum events is realized in some parallel universe.

In summary, the many-worlds interpretation of quantum mechanics provides a quantum mechanical basis for the multiverse concept by positing the existence of a vast ensemble of parallel universes, each corresponding to a different branch of the wave function. This interpretation offers a deterministic approach to quantum physics and eliminates the need for wave function collapse, supporting the idea that all possible outcomes of quantum measurements are physically realized in separate branches of reality.

Philosophical and Conceptual Support

Anthropic Principle: Some physicists use the anthropic principle, which states that observations of the universe must be compatible with the conscious and sapient life that observes it, to support the multiverse. This principle suggests that our universe may be one of many that has the right conditions for life, explaining fine-tuning without invoking intelligent design.

- The anthropic principle is a philosophical and scientific concept that seeks to explain certain observed features of the universe, particularly its apparent fine-tuning for the existence of life. It argues that the universe's properties must be compatible with the existence of conscious and sapient observers, such as humans, because without such observers, the properties would not be observed or known. The anthropic principle is often invoked in discussions about the multiverse to explain why our universe appears finely tuned for life without resorting to notions of intelligent design. Here's how the anthropic principle supports the multiverse:

1. Fine-Tuning and Life-Permitting Conditions: The concept of fine-tuning refers to the remarkable precision and balance of fundamental physical constants and parameters in our universe that allow for the emergence and sustenance of life. Examples of fine-tuning include the strength of gravitational and electromagnetic forces, the mass of elementary particles, and the cosmological constant. The anthropic principle suggests that our universe exhibits these life-permitting conditions because only in such a universe could conscious observers like ourselves emerge and contemplate the universe's properties.

2. Selection Bias and Observer Effect: According to the anthropic principle, the apparent fine-tuning of the universe for life is not surprising or miraculous but rather a consequence of selection bias and the observer effect. Conscious observers can only exist in universes where the conditions are conducive to their existence. Therefore, it is not surprising that we find ourselves in a universe with properties compatible with life, as universes without such properties would not support observers capable of making observations.

3. Multiverse and Anthropic Selection: The multiverse hypothesis provides a natural explanation for the apparent fine-tuning of our universe by invoking a vast ensemble of parallel universes with varying properties. According to the multiverse scenario, each universe in the ensemble has its own set of physical laws and constants, determined by chance or underlying fundamental principles. The anthropic principle then suggests that observers like ourselves can only exist in those universes where the conditions are suitable for life, explaining why our universe appears finely tuned.

4. Avoiding Intelligent Design: By invoking the anthropic principle and the multiverse hypothesis, physicists can explain the apparent fine-tuning of the universe without resorting to notions of intelligent design or cosmic purpose. Instead of positing a deliberate act of creation or design by a supernatural entity, the anthropic principle suggests that our universe's properties are a consequence of natural processes and the existence of multiple universes with varying characteristics.

In summary, the anthropic principle supports the multiverse hypothesis by providing a framework for understanding the apparent fine-tuning of the universe for life. By arguing that conscious observers can only exist in universes with life-permitting conditions, the anthropic principle suggests that our universe may be one of many in a vast multiverse, each with its own set of physical laws and properties. This perspective avoids the need for intelligent design while offering a natural explanation for the observed features of our universe.

Expanding the Scope of Science: Proponents argue that the multiverse theory expands the scope of scientific inquiry, pushing the boundaries of what is considered knowable and observable. This broader perspective may be necessary for future theoretical and technological advancements.

- The multiverse theory represents a significant expansion of the scope of scientific inquiry, challenging traditional notions of what is considered knowable and observable in the universe. Proponents of the multiverse theory argue that embracing this broader perspective is essential for advancing our understanding of the cosmos and unlocking new theoretical and technological frontiers. Here's how the multiverse theory expands the scope of science:

1. Exploring Beyond the Observable Universe: The concept of the multiverse extends scientific inquiry beyond the observable universe, which is limited by the speed of light and the age of the universe. While current observational techniques allow scientists to study distant galaxies and cosmic phenomena, the multiverse hypothesis suggests that there may be regions of space-time beyond our observable horizon that contain other universes with different properties and physical laws. Exploring the multiverse requires new theoretical frameworks and observational techniques capable of probing these inaccessible realms.

2. Investigating Fundamental Principles: The multiverse theory encourages scientists to explore the fundamental principles governing the cosmos on a deeper level. By considering the possibility of multiple universes with varying characteristics, researchers can investigate the underlying mechanisms responsible for generating diverse physical laws and properties. This exploration may lead to insights into the fundamental nature of reality, including the origin of the universe, the nature of space and time, and the fundamental forces of nature.

3. Testing Theoretical Frameworks: The multiverse theory provides a fertile testing ground for theoretical frameworks in physics and cosmology. Scientists can use mathematical models and computational simulations to explore the consequences of different multiverse scenarios and compare them with observational

data. By testing the predictions of multiverse theories against empirical evidence, researchers can refine and improve our understanding of the underlying physics governing the multiverse and our place within it.

4. Stimulating Technological Innovation: The exploration of the multiverse theory may stimulate technological innovation and the development of new observational techniques. Advanced telescopes, particle accelerators, and computational algorithms may be needed to detect indirect evidence of other universes or to probe the fundamental properties of the multiverse. Additionally, interdisciplinary collaborations between physicists, astronomers, mathematicians, and computer scientists may lead to breakthroughs in theoretical and experimental research related to the multiverse.

5. Inspiring Imagination and Creativity: Finally, the multiverse theory inspires imagination and creativity among scientists and the general public alike. By contemplating the existence of other universes with different physical laws and properties, researchers are encouraged to think creatively and explore unconventional ideas. This openness to new possibilities may lead to unexpected discoveries and paradigm shifts in our understanding of the cosmos.

In summary, the multiverse theory expands the scope of scientific inquiry by challenging traditional boundaries and encouraging exploration beyond the observable universe. By investigating the implications of multiple universes, scientists can deepen our understanding of fundamental principles, test theoretical frameworks, stimulate technological innovation, and inspire imagination and creativity in the pursuit of knowledge about the cosmos.

Continuing Debates

Methodological and Epistemological Debates: The discussion around the multiverse theory also includes deeper philosophical debates about the nature of science itself—what constitutes proper scientific methodology, the role of predictions, and the limits of human knowledge.

- The multiverse theory engenders methodological and epistemological debates that delve into fundamental questions about the nature of science, the validity of scientific methodology, the role of predictions, and the boundaries of human knowledge. These debates extend beyond the empirical evidence and mathematical formulations associated with the multiverse hypothesis, touching on broader philosophical issues that shape our understanding of the scientific enterprise. Here's how the multiverse theory contributes to these discussions:

1. Scientific Methodology: The multiverse theory challenges traditional notions of scientific methodology by proposing hypotheses that may be difficult or impossible to test directly through empirical observation. Critics argue that if other universes are unobservable or causally disconnected from our own, the multiverse hypothesis may fall outside the realm of empirical science, which relies on testable predictions and falsifiable hypotheses. Proponents, however, argue that the multiverse theory is rooted in mathematical frameworks consistent with known physical laws and should be considered a legitimate scientific hypothesis despite its observational challenges.

2. Role of Predictions: In discussions about the multiverse theory, questions arise about the role of predictions in scientific inquiry. Some critics argue that if a theory cannot make testable predictions or offer empirical evidence to support its claims, it lacks scientific validity. However, proponents of the multiverse point out that while direct observational evidence of other universes may be elusive, the theory can still make predictions about observable phenomena within our universe that are consistent with multiverse models. These predictions may include features of the cosmic microwave background radiation, the distribution of galaxies, or the behavior of fundamental particles.

3. Limits of Human Knowledge: The multiverse theory prompts reflection on the limits of human knowledge and our ability to understand the universe. Philosophical debates arise about whether certain aspects of reality, such as the existence of other universes, are inherently beyond the scope of human comprehension. Skeptics argue that without empirical evidence or observational access to other universes, claims about the multiverse may amount to speculative metaphysics rather than empirical science. Proponents counter that scientific progress often involves exploring concepts that initially lie beyond our direct observation, such as subatomic particles or distant galaxies, and that the multiverse hypothesis is a natural extension of this process.

4. Philosophical Implications: Beyond its scientific implications, the multiverse theory raises profound philosophical questions about the nature of existence, reality, and our place within the cosmos. Discussions revolve around issues of ontology—what entities exist and how they are defined—as well as the nature of causality,

contingency, and determinism in a multiverse frame-work. Philosophers also explore the implications of the multiverse for questions of meaning, purpose, and the search for ultimate explanations about the universe's origin and structure.

In summary, the discussion surrounding the multiverse theory extends beyond scientific evidence to encompass methodological and epistemological debates about the nature of science, the role of predictions, the limits of human knowl-edge, and the philosophical implications of multiverse con-cepts. These debates highlight the interdisciplinary nature of the multiverse discourse and its significance for under-standing both the scientific and philosophical dimensions of reality.

Emerging Technologies and Future Tests: As technology ad-vances, new methods may emerge that could potentially provide indirect evidence of the multiverse, such as precise measurements of cosmic background radiation or detecting signs of gravitational effects from other universes.

- Emerging technologies hold the promise of enabling novel methods for investigating the multiverse hypothesis, offering potential avenues for obtaining indirect evidence or obser-vational signatures that could support or refute multiverse models. As scientific instrumentation becomes more sophis-ticated and computational capabilities increase, researchers are exploring innovative approaches to probe the cosmos for clues about the existence of other universes. Here are some emerging technologies and future tests that could shed light on the multiverse hypothesis:

1. Precision Cosmology: Advancements in precision cosmology, particularly in the measurement of the cosmic microwave background (CMB) radiation, offer opportunities to glean insights into the structure and evolution of the universe. Future experiments such as the Cosmic Microwave Background Stage-4 (CMB-S4) survey aim to map the CMB with unprecedented detail, potentially revealing subtle signatures indicative of a multiverse scenario. By studying fluctuations in the CMB temperature and polarization across different angular scales, scientists may detect patterns or anomalies that could be interpreted as evidence of other universes imprinting their influence on our observable cosmos.

2. Gravitational Wave Astronomy: The recent detection of gravitational waves by experiments like the Laser Interferometer Gravitational-Wave Observatory (LIGO) and the Virgo Collaboration has opened up a new window for exploring the universe. Future generations of gravitational wave detectors, such as space-based interferometers like LISA (Laser Interferometer Space Antenna), could provide insights into the existence of other universes through the detection of gravitational effects propagating from neighboring regions of the multiverse. If other universes exert gravitational influences on ours, these effects might manifest as subtle deviations or correlations in the observed gravitational wave signals.

3. Multimessenger Astronomy: Multimessenger astronomy, which combines observations from different cosmic messengers such as electromagnetic radiation, cosmic rays, neutrinos, and gravitational waves, holds promise for uncovering hidden facets of the universe. By correlating data from multiple sources, scientists

may discern signatures of the multiverse that are not apparent in individual datasets alone. For example, combining observations of high-energy cosmic rays with gravitational wave signals or CMB measurements could provide complementary information about the underlying structure of the multiverse.

4. Quantum Computing and Simulation: Advances in quantum computing and simulation present opportunities for modeling and simulating complex multiverse scenarios with greater fidelity and efficiency. Quantum computers could be used to explore the behavior of multiverse models derived from string theory or inflationary cosmology, allowing researchers to simulate the evolution of multiple universes and test theoretical predictions. By comparing simulated multiverse scenarios with observational data, scientists may gain insights into the feasibility and implications of different multiverse hypotheses.

5. Astroparticle Physics: Experiments in astroparticle physics, such as those studying high-energy cosmic rays, dark matter, and neutrinos, could provide indirect evidence of the multiverse through anomalous signatures or unexpected phenomena. By analyzing cosmic ray spectra, for example, researchers may detect unusual patterns or distributions that could be attributed to interactions with particles or fields originating from other universes. Similarly, studies of dark matter properties or neutrino oscillations might reveal subtle deviations consistent with multiverse predictions.

In summary, emerging technologies offer promising avenues for future tests of the multiverse hypothesis, providing opportunities to detect indirect evidence or observational signatures that could illuminate the existence of other universes. From precision cosmology and gravitational wave astronomy to quantum computing and astroparticle physics, these advancements hold the potential to revolutionize our understanding of the cosmos and unlock the secrets of the multiverse.

Conclusion

The debate over the multiverse theory encapsulates a fundamental scientific dialectic—between skepticism rooted in empirical traditions and the speculative reach of theoretical physics. As our tools and theories evolve, so too will our understanding of the universe (or universes). Whether the multiverse theory will move from the peripheries of speculative science to a more empirically grounded theory depends on both the ingenuity of theorists and the evolution of observational technologies.

In the next chapter, we will explore the ethical and moral implications of the multiverse theory, considering how it challenges or reinforces our views on destiny, morality, and the nature of human existence.

~ 10 ~

THE ETHICS OF KNOWING
OTHER REALITIES

In this chapter, we explore the ethical and moral implications of the multiverse theory. The notion that multiple, possibly infinite, universes exist alongside our own raises profound ethical questions about responsibility, choice, and our understanding of existence. We'll examine how these concepts might alter our philosophical outlook on life and the ethical frameworks we traditionally employ.

Redefining Free Will and Moral Responsibility

Multiplicity of Choices: In a multiverse where every possible outcome of a decision exists in some universe, the concept of free will encounters complex new dimensions. If every choice is realized somewhere, what does it mean to make a decision? This chapter delves into the implications for moral responsibility when actions have infinite repercussions.

- The concept of the multiplicity of choices within a multiverse introduces profound implications for the nature of decision-making, free will, and moral responsibility. In a multiverse scenario where every possible outcome of a decision is realized in some universe, the concept of choice takes on new dimensions, challenging traditional understandings of agency and determinism. Here's an exploration of the implications of the multiplicity of choices within a multiverse framework:

1. Agency and Determinism: The existence of multiple universes, each representing a different outcome of a decision, raises questions about the nature of agency and determinism. On one hand, individuals may feel a sense of empowerment knowing that every choice they make has the potential to shape their personal trajectory within the multiverse. However, the multiplicity of choices also suggests that the outcomes of decisions may be predetermined to some extent, as every possible outcome is already realized in other universes. This tension between agency and determinism complicates our understanding of free will within a multiverse context.

2. Moral Responsibility: The multiplicity of choices within a multiverse poses challenges to traditional notions of moral responsibility. If every possible action is realized in some universe, individuals may question the significance of their moral decisions and the consequences of their actions. For example, if one chooses to act morally in one universe but learns that an alternate version of themselves commits an immoral act in another universe, does their moral responsibility extend across multiple realities? The existence of parallel universes blurs the boundaries of moral culpability and accountability, raising complex ethical dilemmas.

3. Ethical Frameworks: The multiplicity of choices within a multiverse challenges existing ethical frameworks that rely on concepts of causality, intentionality,

and accountability. Traditional ethical theories, such as consequentialism, deontology, and virtue ethics, may need to be reevaluated in light of the multiverse hypothesis. Ethicists may need to develop new frameworks that account for the interconnectedness of choices across multiple realities and the implications for moral judgment and decision-making.

4. Existential Reflection: The concept of the multiplicity of choices within a multiverse invites existential reflection on the nature of identity, selfhood, and the meaning of life. Individuals may grapple with questions about the significance of their actions in the vastness of the multiverse and the role of choice in shaping their personal narrative across multiple realities. Existential angst may arise as individuals confront the infinite possibilities and uncertainties inherent in a multiverse worldview, prompting them to reassess their values, priorities, and existential goals.

5. Philosophical Inquiry: The multiplicity of choices within a multiverse stimulates philosophical inquiry into the nature of reality, causality, and consciousness. Philosophers may explore theories of modal realism, branching spacetime, and quantum indeterminacy to better understand the implications of the multiverse for our conceptual frameworks. Questions about the ontology of parallel universes, the nature of identity across realities, and the relationship between consciousness and existence become central topics of philosophical debate within a multiverse context.

In summary, the concept of the multiplicity of choices within a multiverse introduces complex philosophical questions about agency, determinism, moral responsibility, existential reflection, and philosophical inquiry. Exploring these implications enriches our understanding of the multiverse hypothesis and its broader implications for our conception of reality and the human experience.

Impact on Ethics: Traditional ethical theories often hinge on the consequences of our actions. The multiverse theory complicates this by suggesting that every conceivable consequence happens somewhere. We explore how this might affect ethical decision-making, particularly whether it diminishes the weight of our choices or amplifies our responsibility to choose wisely.

- The impact of the multiverse theory on ethics is profound, as it challenges traditional ethical frameworks by introducing complexities related to the consequences of our actions across multiple realities. Here's an exploration of how the multiverse theory may affect ethical decision-making and the weight of our choices:

1. Consequentialism and Utilitarianism: Consequentialist ethical theories, such as utilitarianism, judge the morality of actions based on their outcomes and the overall happiness or well-being they produce. However, in a multiverse where every conceivable consequence of an action occurs in some universe, the idea of maximizing utility or minimizing harm becomes more nuanced. The realization that all possible outcomes are realized somewhere challenges the notion of quantifying or comparing the consequences of actions across multiple realities. Ethicists may need to reconsider how they evaluate the ethical implications of actions in light of the multiverse hypothesis.

2. Moral Luck and Responsibility: The multiverse theory introduces the concept of moral luck, whereby the moral status of an action depends on factors beyond an individual's control, such as which universe their actions occur in and the resulting consequences. Individuals may grapple with questions about the extent of their moral responsibility when the outcomes of their actions are predetermined by the existence of parallel universes. Does the fact that an alternate version of oneself commits a harmful act in another universe diminish the moral culpability of one's own actions? Alternatively, does the knowledge that every choice has infinite repercussions amplify the moral responsibility to choose wisely and act ethically?

3. Existential Reflection and Meaning: The multiverse theory prompts existential reflection on the significance of our choices and actions in the vastness of the multiverse. Individuals may question the meaning of morality and the value of ethical decision-making when confronted with the realization that every possible consequence of their actions is already realized somewhere. This existential angst may lead to a reevaluation of one's values, priorities, and ethical commitments in light of the multiplicity of choices within a multiverse.

4. Ethical Deliberation and Decision-Making: Ethical decision-making becomes more complex in a multiverse framework, as individuals must navigate the moral implications of their actions across infinite realities. The uncertainty surrounding the consequences of actions may lead to ethical paralysis or decision-making fatigue, as individuals grapple with the impossibility of predicting or controlling the outcomes of their choices. Ethicists may need to develop new approaches to ethical deliberation that account for the

interconnectedness of choices across multiple realities and the inherent uncertainty of moral decision-making within a multiverse context.

5.   Ethical Relativism and Pluralism: The multiverse theory challenges notions of ethical relativism and pluralism by suggesting that all possible moral codes and value systems are realized in some universe. This realization raises questions about the universality of ethical principles and the relativity of moral judgments across different realities. Ethicists may need to explore how ethical norms and values vary across parallel universes and how individuals navigate ethical dilemmas in contexts where moral frameworks differ.

In summary, the multiverse theory complicates traditional ethical theories by introducing uncertainties and complexities related to the consequences of our actions across multiple realities. Ethical decision-making becomes more nuanced as individuals grapple with questions about moral responsibility, existential reflection, and the significance of their choices in a multiverse framework. Exploring these implications enriches our understanding of ethics and the human experience within the vastness of the multiverse.

Cultural and Social Implications

Cultural Relativism in a Multiversal Context: Understanding that different universes could operate under entirely different moral and physical laws might lead to a form of "multiversal relativism," where the ethics of one universe aren't easily applicable to another. This section discusses how such a perspective might influence global ethics and international relations.

- Cultural relativism in a multiversal context introduces the concept of "multiversal relativism," which acknowledges that different universes may operate under entirely distinct moral and physical laws. This perspective challenges the universality of ethical principles and suggests that moral frameworks are contingent on the specific conditions of each universe. Here's an exploration of how multiversal relativism might influence global ethics and international relations:

1. Pluralism and Diversity: Multiversal relativism recognizes the diversity of moral norms and values across different universes, highlighting the plurality of ethical frameworks that exist within the multiverse. This perspective encourages a more inclusive and tolerant approach to cultural differences, as it acknowledges that what is considered morally acceptable in one universe may not be applicable or relevant in another. Embracing multiversal relativism promotes respect for cultural diversity and fosters dialogue among civilizations with divergent ethical perspectives.

2. Moral Universes and Interactions: In a multiversal context, each universe can be seen as a distinct "moral universe" with its own set of ethical principles and values. Multiversal relativism suggests that interactions between universes may involve negotiating differences in moral frameworks and navigating ethical dilemmas that arise from conflicting norms. International relations in a multiverse characterized by multiversal relativism require diplomats and policymakers to adopt a diplomatic approach that respects the sovereignty of each universe while promoting dialogue and cooperation on shared challenges.

3. Ethical Exchange and Learning: Multiversal relativism opens up opportunities for ethical exchange and

mutual learning among civilizations inhabiting different universes. Just as cultures on Earth engage in cross-cultural dialogue and exchange, civilizations in a multiverse can share insights, perspectives, and ethical practices with one another. This exchange enriches the ethical discourse and contributes to the development of a more nuanced understanding of morality within the multiverse. Diplomatic initiatives aimed at fostering ethical dialogue and collaboration can promote mutual understanding and cooperation across universes.

4. Ethical Challenges and Conflicts: While multiversal relativism promotes tolerance and understanding of cultural diversity, it also poses challenges in addressing ethical conflicts and human rights abuses that transcend universes. In situations where moral principles clash or where one universe imposes its values on another, diplomatic efforts must navigate complex ethical terrain to uphold universal human rights and promote peace and stability. International organizations and inter-universal alliances may play a crucial role in mediating disputes and resolving conflicts in a multiverse characterized by multiversal relativism.

5. Ethical Frameworks and Multiversal Governance: Multiversal relativism raises questions about the feasibility of establishing universal ethical frameworks or governance structures that apply across all universes. While efforts to promote ethical standards and norms may have merit, the diversity of moral perspectives within the multiverse necessitates a flexible and inclusive approach to governance. Multiversal governance initiatives must accommodate the multiplicity of ethical frameworks and prioritize consensus-building and dialogue among universes to address common challenges and uphold shared values.

In summary, cultural relativism in a multiversal context gives rise to the concept of multiversal relativism, which acknowledges the diversity of moral frameworks across different universes. This perspective influences global ethics and international relations by promoting pluralism, fostering ethical exchange and learning, addressing ethical challenges and conflicts, and shaping ethical frameworks and governance structures within the multiverse. Embracing multiversal relativism encourages a more inclusive and collaborative approach to navigating the complexities of morality across multiple realities.

Social Justice and Parallel Outcomes: Consideration of how different life circumstances in parallel universes might affect our views on fairness and equality in our own universe. If every person potentially experiences every possible outcome, how does this change our approach to social justice and equality?

- The concept of social justice and parallel outcomes within a multiverse framework prompts a reevaluation of fairness, equality, and our understanding of the distribution of opportunities and resources across different universes. Here's an exploration of how consideration of parallel outcomes might influence our approach to social justice and equality:

1. Intersectionality and Multiple Realities: In a multiverse where every person potentially experiences every possible outcome, the concept of intersectionality takes on new dimensions. Intersectionality recognizes that individuals occupy multiple social positions and identities that intersect to shape their experiences and opportunities. In a multiverse context, individuals

may navigate different social contexts, identities, and life circumstances across parallel universes, leading to a more nuanced understanding of privilege, disadvantage, and systemic oppression. Social justice movements must account for the multiplicity of realities within the multiverse and address intersecting forms of discrimination and marginalization.

2. Redistribution and Resource Allocation: The existence of parallel outcomes challenges traditional notions of meritocracy and individual responsibility by highlighting the role of chance and circumstance in shaping life trajectories. In a multiverse where every person potentially experiences every possible outcome, the distribution of resources and opportunities becomes more complex. Social justice advocates may argue for more equitable forms of redistribution and resource allocation that acknowledge the role of systemic factors and structural barriers in determining life outcomes across parallel universes. Policies aimed at addressing inequalities must take into account the multiplicity of realities within the multiverse and prioritize justice and fairness for all individuals, regardless of the universe they inhabit.

3. Ethical Considerations and Moral Responsibility: The concept of parallel outcomes raises ethical considerations about moral responsibility and collective action within the multiverse. Individuals may grapple with questions about their obligations to address injustices that occur in parallel universes and the extent to which they are morally responsible for the outcomes experienced by alternate versions of themselves. Social justice movements may advocate for a more expansive understanding of moral responsibility that extends beyond individual actions to encompass systemic factors and

collective efforts to address inequalities across multiple realities.

4.   Empathy and Solidarity: Understanding the multiplicity of outcomes within a multiverse fosters empathy and solidarity across diverse experiences and life circumstances. Individuals may develop a greater appreciation for the interconnectedness of human experiences and the shared struggle for justice and equality across parallel universes. Social justice movements may emphasize the importance of empathy and solidarity as guiding principles in advocating for systemic change and challenging injustices that transcend universes. Building alliances and coalitions that bridge differences and unite individuals across parallel realities strengthens collective efforts to promote social justice and equality within the multiverse.

5.   Reimagining Systems and Institutions: The existence of parallel outcomes invites a reimagining of systems and institutions to better accommodate the diversity of human experiences and life trajectories across the multiverse. Social justice advocates may propose transformative reforms that prioritize inclusivity, accessibility, and equity in designing policies and practices that impact individuals across parallel universes. By centering the voices and experiences of marginalized communities and recognizing the inherent dignity and worth of all individuals, societies can work towards creating more just and equitable systems that reflect the values of solidarity and shared humanity within the multiverse.

In summary, consideration of parallel outcomes within a multiverse framework prompts a reevaluation of social justice and equality by highlighting the interconnectedness of human experiences and the importance of addressing systemic injustices across diverse realities. By acknowledging the multiplicity of outcomes and advocating for transformative change, individuals and societies can work towards creating more just and equitable systems that reflect the values of empathy, solidarity, and collective responsibility within the multiverse.

Theological and Metaphysical Implications

Theology Across Universes: Different religious and spiritual beliefs offer various interpretations of creation, existence, and morality. The idea of a multiverse might challenge these beliefs or require adaptations of theological doctrines to encompass multiple, possibly contradictory, realities.

- The concept of a multiverse presents intriguing challenges and opportunities for theology, as it prompts religious and spiritual traditions to grapple with the implications of multiple, possibly contradictory, realities. Here's an exploration of how theology across universes might evolve in response to the idea of a multiverse:

1. Cosmological Interpretations: Many religious and spiritual traditions offer cosmological narratives that explain the origins and nature of the universe(s). The concept of a multiverse challenges traditional cosmological interpretations by suggesting the existence of multiple, parallel realities with potentially different physical laws and properties. Theological reflections on the multiverse may involve reinterpreting creation myths and cosmological narratives to accommodate

the idea of multiple universes and their relationship to divine creation.

2.   Divine Sovereignty and Omnipotence: The existence of a multiverse raises questions about divine sovereignty and omnipotence within theological frameworks. If there are multiple universes with different physical laws and outcomes, how does this affect traditional understandings of God's control over creation? Theological reflections on the multiverse may explore the compatibility of divine sovereignty with the existence of diverse realities and the role of God in sustaining and governing multiple universes.

3.   Plurality of Religious Experience: The concept of a multiverse suggests the possibility of diverse religious experiences and spiritual paths across parallel realities. Theological reflections on the multiverse may acknowledge the plurality of religious traditions and spiritual practices within and across universes, emphasizing the importance of respecting and learning from different belief systems. This perspective invites theologians to explore the theological implications of inter-universal dialogue and the mutual enrichment of religious traditions in a multiverse context.

4.   Moral and Ethical Considerations: The idea of a multiverse challenges theological doctrines related to morality and ethics by introducing the possibility of alternate moral frameworks and outcomes across different realities. Theological reflections on the multiverse may explore the implications of moral relativism and universalism within and across universes, considering how ethical principles and divine commandments apply across diverse realities. This exploration may lead to theological adaptations that emphasize the

universality of moral truths or the contextual nature of ethical norms within a multiverse framework.

5.   Eschatological Perspectives: Many religious traditions offer eschatological beliefs about the ultimate destiny and purpose of creation. The concept of a multiverse invites theological reflections on the eschatological implications of multiple realities and the interconnectedness of cosmic destinies across universes. Theological perspectives on the multiverse may explore themes of cosmic redemption, universal salvation, or the reconciliation of diverse realities in divine providence.

6.   Mystical and Spiritual Insights: Some mystical and spiritual traditions offer insights into the nature of reality that transcend conventional cosmological frameworks. Theological reflections on the multiverse may draw on mystical experiences and spiritual teachings to explore the interconnectedness of all existence within the multiverse. This perspective may enrich theological discourse by highlighting the unity of all creation across diverse realities and the transformative potential of spiritual awakening in a multiverse context.

In summary, the concept of a multiverse challenges theology to adapt and evolve in response to the complexities of multiple, possibly contradictory, realities. Theological reflections on the multiverse may explore cosmological interpretations, divine sovereignty, religious pluralism, moral and ethical considerations, eschatological perspectives, and mystical insights to deepen understanding of the divine within the multiverse. By engaging with the theological implications of the multiverse, religious and spiritual traditions can enrich their perspectives on creation, existence, and morality across diverse realities.

Metaphysical Responsibility: This section considers whether knowledge of the multiverse imposes new kinds of metaphysical duties—responsibilities not just to the observable world but to the entirety of existence across universes.

- The concept of metaphysical responsibility in the context of the multiverse raises intriguing questions about our ethical obligations and moral duties beyond the confines of our observable universe. Here's an exploration of how knowledge of the multiverse may impose new kinds of metaphysical duties:

  1. Stewardship of Existence: Knowledge of the multiverse expands our conception of existence beyond the boundaries of our observable universe. Metaphysical responsibility entails a duty to act as stewards of the entirety of existence across universes, acknowledging our interconnectedness with all forms of reality. This stewardship may involve ethical considerations related to the preservation, flourishing, and harmonious co-existence of diverse realities within the multiverse.

  2. Care for Alternate Realities: The existence of multiple, possibly infinite, universes raises questions about our ethical obligations towards alternate realities and their inhabitants. Metaphysical responsibility requires us to consider the well-being and moral significance of beings and phenomena across parallel universes, even if they are beyond our immediate perception. This care for alternate realities may involve promoting justice, compassion, and flourishing across diverse realms within the multiverse.

3. Respect for Cosmic Order: Metaphysical responsibility entails a respect for the cosmic order and the underlying principles that govern existence across universes. This includes recognizing the interconnectedness of all phenomena within the multiverse and acting in accordance with principles of cosmic harmony and balance. Ethical considerations related to metaphysical responsibility may involve fostering reverence for the intrinsic value of all aspects of reality and upholding the integrity of the cosmic order across diverse universes.

4. Ethical Decision-Making Across Universes: Knowledge of the multiverse raises ethical questions about decision-making and actions that have consequences beyond our own universe. Metaphysical responsibility requires us to consider the broader implications of our choices and behaviors across multiple realities, recognizing that our actions may have far-reaching effects on beings and phenomena in other universes. This ethical decision-making process may involve deliberating on the ethical implications of inter-universal interactions and striving to act in ways that promote the well-being and integrity of existence across the multiverse.

5. Exploration and Understanding: Metaphysical responsibility includes a commitment to exploration and understanding of the multiverse and its manifold dimensions. This entails cultivating curiosity, wisdom, and humility in our efforts to comprehend the complexities of existence across universes and to appreciate the diversity and richness of reality. Ethical considerations related to metaphysical responsibility may involve fostering a spirit of inquiry, collaboration, and mutual respect in our exploration of the multiverse and our engagement with beings and phenomena beyond our own universe.

In summary, metaphysical responsibility in the context of the multiverse encompasses a duty to act as stewards of existence, care for alternate realities, respect the cosmic order, engage in ethical decision-making across universes, and foster exploration and understanding of the multiverse. By embracing our interconnectedness with all forms of reality and acknowledging the ethical implications of our actions beyond our observable universe, we can fulfill our metaphysical duties and contribute to the flourishing and harmony of existence across the vast expanse of the multiverse.

Philosophical Perspectives on Identity and Existence

The Self Across Multiverses: Philosophical inquiry into whether a coherent self exists across multiverses. Does knowledge of potential other 'selves' alter our perception of identity and selfhood? This discussion draws on existential and phenomenological philosophy to examine the nature of self in a multiversal context.

- Exploring the concept of the self across multiverses involves delving into existential and phenomenological philosophy to contemplate the nature of identity and selfhood in the context of multiple, possibly infinite, parallel realities. Here's an exploration of how knowledge of potential other 'selves' may alter our perception of identity:

1. Existential Perspective: Existential philosophy delves into questions of existence, freedom, and the nature of the self. In a multiversal context, existential inquiry may examine how the awareness of multiple potential 'selves' across universes impacts our sense of individuality and authenticity. The existential notion

of 'thrownness'—the idea that we find ourselves in a world not of our own making—takes on new dimensions when considering the multiplicity of selves across multiverses. Existentialists may question whether there is a core, immutable self that persists across diverse realities or whether identity is fluid and contingent upon the particularities of each universe.

2.   Phenomenological Inquiry: Phenomenology focuses on the study of subjective experience and the structures of consciousness. In the context of multiverses, phenomenological analysis may explore how the perception of alternate 'selves' influences our lived experience and self-awareness. Phenomenologists may investigate the intersubjective dimensions of identity, considering how the encounter with other potential 'selves' in parallel realities shapes our understanding of selfhood. Phenomenological reflections may reveal the interplay between the 'I' and the 'other' across multiverses, shedding light on the relational nature of identity and the multiplicity of selves within consciousness.

3.   Fluidity of Identity: Knowledge of potential other 'selves' across multiverses challenges traditional conceptions of identity as singular and coherent. Existential and phenomenological perspectives may highlight the fluidity and plurality of identity, suggesting that the self is not a fixed entity but a dynamic process that unfolds across diverse realities. This fluidity of identity may manifest as a sense of existential vertigo—a disorienting awareness of the myriad possibilities and permutations of selfhood across multiverses.

4.   Authenticity and Choice: Existential philosophy emphasizes the importance of authenticity—the existential commitment to self-awareness, freedom, and responsibility. In a multiversal context, questions of

authenticity become nuanced as individuals confront the multiplicity of potential 'selves' and the existential dilemmas inherent in navigating diverse realities. Existentialists may explore how the awareness of alternate 'selves' influences our choices, values, and sense of purpose, as well as the existential anxiety that arises from the recognition of infinite possibilities and the uncertainty of identity.

5.   Integration and Transcendence: Existential and phenomenological reflections on the self across multiverses may culminate in an invitation to integrate and transcend the multiplicity of potential 'selves' within a coherent framework of identity. This process involves reconciling the diversity of experiences and perspectives across universes while affirming a fundamental sense of selfhood that transcends individual manifestations. Existentialists may advocate for embracing the existential paradox of being both finite and infinite, individual and universal, as we navigate the multiversal landscape of existence.

In summary, philosophical inquiry into the self across multiverses draws on existential and phenomenological perspectives to examine the nature of identity and selfhood in the context of multiple potential 'selves' across parallel realities. By grappling with questions of authenticity, choice, and integration, individuals may cultivate a deeper understanding of their existential condition and the existential significance of the multiverse.

Existential Implications: How does the multiverse theory affect existential questions about meaning and purpose? If all outcomes are realized, does it enhance or diminish the search for meaning in life?

- The multiverse theory introduces profound existential implications, challenging traditional perspectives on meaning and purpose by suggesting that all possible outcomes are realized across parallel universes. Here's an exploration of how the multiverse theory affects existential questions about meaning and purpose:

1. Amplification of Existential Angst: The multiverse theory may amplify existential angst—the profound sense of disorientation and anxiety that arises from the awareness of infinite possibilities and the uncertainty of existence. If all outcomes are realized somewhere in the multiverse, individuals may grapple with feelings of insignificance and existential vertigo as they confront the vastness and complexity of reality. This amplification of existential angst may intensify the search for meaning as individuals seek to navigate the multiversal landscape of existence.

2. Expansion of Possibilities: On the other hand, the multiverse theory expands the scope of possibilities for meaning and purpose, offering individuals a broader canvas upon which to explore their existential quests. If every conceivable outcome is actualized somewhere in the multiverse, individuals may find solace in the idea that their choices and actions have significance across diverse realities. This expansion of possibilities may inspire a sense of existential liberation as individuals embrace the multiplicity of potential paths and outcomes in their search for meaning.

3. Pluralism of Meaning: The multiverse theory fosters a pluralistic understanding of meaning—a recognition that meaning is not fixed or universal but contingent upon the particularities of each universe and individual experience. In a multiversal context, meaning may manifest as a mosaic of diverse narratives, values, and purposes that emerge from the interactions between different 'selves' and realities. This pluralism of meaning invites individuals to explore the richness and complexity of existence across parallel universes, embracing the diversity of existential possibilities.

4. Reevaluation of Purpose: The multiverse theory prompts a reevaluation of traditional notions of purpose and teleology—the idea that there is a predetermined end or goal to existence. If all outcomes are realized in the multiverse, individuals may question whether there is a singular, overarching purpose to life or whether purpose emerges from the interplay of choices and experiences across diverse realities. This reevaluation of purpose encourages individuals to adopt a more fluid and dynamic understanding of meaning, embracing the ongoing process of existential exploration and self-discovery.

5. Integration of Perspectives: Ultimately, the multiverse theory invites individuals to integrate diverse perspectives on meaning and purpose, recognizing that existential questions are inherently complex and multifaceted. Rather than seeking definitive answers to questions of meaning, individuals may embrace the journey of existential inquiry, drawing inspiration from the plurality of narratives and experiences across the multiverse. This integration of perspectives fosters a more holistic and inclusive approach to existential

exploration, enriching individuals' understanding of the profound mysteries of existence.

In summary, the multiverse theory affects existential questions about meaning and purpose by amplifying existential angst, expanding the scope of possibilities, fostering a pluralistic understanding of meaning, prompting a reevaluation of purpose, and inviting the integration of diverse perspectives. By grappling with the existential implications of the multiverse, individuals may embark on a journey of profound self-discovery and existential exploration, enriching their understanding of the complex tapestry of existence across parallel universes.

Ethical Considerations in Multiverse Research

Ethics of Multiverse Research: As scientists push the boundaries of what can be known and explored, ethical considerations in conducting multiverse research are paramount. This includes the responsibility of scientists to communicate their findings without causing undue alarm or misunderstanding about the nature of reality.

- The ethics of multiverse research encompass a range of considerations that are crucial for scientists as they push the boundaries of knowledge and explore the implications of parallel universes. Here's an expansion on these ethical considerations:

1. Transparency and Communication: Scientists engaged in multiverse research have an ethical responsibility to communicate their findings transparently and

accurately to the public. This includes providing clear explanations of the theoretical concepts involved, the methods employed in research, and the potential implications of the findings. Transparent communication helps to build trust between scientists and the public, ensuring that information about the nature of reality is disseminated responsibly and without sensationalism.

2. Avoiding Misinterpretation: Given the speculative nature of multiverse theories, scientists must exercise caution to prevent their research from being misinterpreted or misrepresented. It's essential to convey that multiverse theories are theoretical frameworks based on mathematical models and scientific principles, rather than proven facts about the nature of reality. By emphasizing the provisional nature of scientific knowledge and the ongoing process of inquiry, scientists can mitigate the risk of misunderstanding and undue alarm among the public.

3. Ethical Use of Resources: Multiverse research often requires significant resources, including funding, equipment, and personnel. Scientists have an ethical obligation to use these resources responsibly and efficiently, ensuring that research efforts contribute to the advancement of knowledge in a meaningful way. This involves prioritizing research projects that have the potential to yield valuable insights into the multiverse theory while considering the broader societal impact of scientific endeavors.

4. Consideration of Societal Impact: Multiverse research may have societal implications that extend beyond the scientific community. Scientists should consider the potential ethical, social, and cultural ramifications of their work, including how it may influence public perceptions, beliefs, and values. By engaging in

dialogue with diverse stakeholders, including ethicists, policymakers, and community members, scientists can better anticipate and address concerns about the ethical implications of multiverse research.

5.   Responsible Speculation: While speculation is inherent in theoretical physics, scientists must exercise caution when discussing speculative concepts such as the multiverse. Ethical multiverse research involves distinguishing between well-supported scientific hypotheses and speculative conjecture, clearly articulating the limitations and uncertainties of theoretical frameworks. Responsible speculation acknowledges the speculative nature of multiverse theories while maintaining scientific rigor and intellectual integrity.

6.   Promoting Ethical Engagement: Scientists have a responsibility to promote ethical engagement with multiverse research among their peers and the broader scientific community. This includes fostering a culture of integrity, transparency, and accountability in scientific inquiry, as well as advocating for ethical standards and guidelines in multiverse research. By promoting ethical conduct and responsible communication, scientists can uphold the integrity of multiverse research and its societal impact.

In summary, the ethics of multiverse research encompass transparency and communication, avoiding misinterpretation, ethical use of resources, consideration of societal impact, responsible speculation, and promoting ethical engagement. By adhering to these ethical principles, scientists can navigate the complexities of multiverse research responsibly and contribute to the advancement of knowledge with integrity and accountability.

Future Scenarios and Precautionary Principles: Looking ahead, we consider potential future discoveries about the multiverse and how they should be ethically approached. The precautionary principle suggests that in the absence of clear data about the effects of new knowledge or technology, steps should be taken to prevent harm. How might this apply to potentially world-altering multiverse discoveries?

- When considering future scenarios and potential discoveries about the multiverse, ethical considerations are paramount, and the precautionary principle offers valuable guidance in navigating uncertain terrain. Here's an exploration of how the precautionary principle might apply to potentially world-altering multiverse discoveries:

1. Anticipating Unforeseen Consequences: One of the central tenets of the precautionary principle is the recognition of uncertainty and the acknowledgment that new knowledge or technologies may have unforeseen consequences. In the context of multiverse discoveries, scientists must anticipate the potential societal, cultural, and ethical implications of revealing knowledge about parallel universes. This includes considering how such revelations might impact individuals' worldviews, beliefs, and values, as well as broader societal structures and institutions.

2. Assessing Risks and Benefits: The precautionary principle emphasizes the need to weigh the potential risks and benefits of new knowledge or technologies before proceeding. In the case of multiverse discoveries, scientists must carefully assess the potential risks associated with revealing information about

parallel universes, including the destabilization of societal norms, ethical dilemmas, and existential angst. At the same time, scientists should consider the potential benefits of advancing our understanding of the cosmos and the potential for transformative insights into the nature of reality.

3. Precautionary Measures: Applying the precautionary principle to multiverse discoveries may involve taking proactive measures to mitigate potential risks and uncertainties. This could include implementing safeguards to ensure responsible communication of findings, engaging in public dialogue and education to promote informed decision-making, and establishing ethical guidelines and regulatory frameworks to govern the ethical conduct of multiverse research. By taking precautionary measures, scientists can help minimize the potential for harm while maximizing the potential benefits of multiverse discoveries.

4. Ethical Deliberation and Engagement: The precautionary principle underscores the importance of ethical deliberation and engagement with stakeholders in decision-making processes. In the context of multiverse discoveries, scientists should actively engage with ethicists, policymakers, community members, and other stakeholders to solicit diverse perspectives, identify potential risks, and develop strategies for responsible research and communication. By fostering open dialogue and collaboration, scientists can ensure that multiverse discoveries are approached ethically and transparently.

5. Adaptive Management: Finally, the precautionary principle encourages adaptive management approaches that allow for ongoing monitoring and reassessment of risks and benefits as new knowledge emerges. In the

dynamic landscape of multiverse research, scientists must remain vigilant to changing circumstances and be prepared to adjust their approaches in response to new information and evolving ethical considerations. This adaptive approach enables scientists to navigate the complexities of multiverse discoveries responsibly and ethically.

In summary, applying the precautionary principle to potentially world-altering multiverse discoveries involves anticipating unforeseen consequences, assessing risks and benefits, implementing precautionary measures, engaging in ethical deliberation and engagement, and adopting adaptive management approaches. By adhering to these principles, scientists can navigate the ethical complexities of multiverse research and contribute to the responsible advancement of knowledge about the cosmos.

Conclusion

The exploration of the multiverse theory brings to light new ethical, cultural, and metaphysical questions that challenge our traditional ways of thinking about morality, existence, and the universe. As we expand our understanding of the cosmos, we must also expand our ethical horizons, considering not just the world we live in but also the many worlds that might be. In the next chapter, we will return to a discussion of specific dimensions explored within the multiverse, examining the unique physical and ethical laws that might govern these realms.

PART TWO

Exploring Individual Dimensions

Welcome to Part Two of *Parallel Lives, Parallel Realities: Unveiling the Multiverse.* Having established a solid foundation in the theories and concepts that support the existence of parallel universes in Part One, we now embark on a voyage of imagination and exploration into the varied dimensions that might exist beyond our known universe. Each chapter in this part will delve into a unique dimension, offering a detailed and imaginative portrayal of its distinct characteristics, the life forms it could support, and the physical laws that govern it.

Purpose and Approach

The chapters ahead are crafted to stretch the boundaries of our understanding and to challenge our perceptions of what is possible. While grounded in the theoretical physics discussed previously, these explorations are speculative by nature, combining scientific insights with creative extrapolation to paint vivid pictures of alternative realities. Our journey through these dimensions is not just an exercise in creativity but also a deeper inquiry into the nature of reality itself.

Themes and Exploration

1. **Diverse Realities**: We will explore dimensions where fundamental aspects such as time, space, gravity, and light behave differently. From worlds where time flows backward to realms composed entirely of sentient flora, each dimension offers a unique scenario that challenges conventional wisdom and scientific norms.

2. **Life and Society**: How would intelligent beings evolve and societies develop under different cosmic rules? Each chapter will speculate on the adaptations of life forms to their environments and the types of social structures that might arise, providing a comprehensive look at life in each proposed universe.

3. **Philosophical and Ethical Implications**: Beyond the scientific and fantastical, we delve into the philosophical ramifications of these alternate dimensions. What can they teach us about our own universe? How do they reflect or invert our understandings of morality, existence, and consciousness?

4. **Interconnectivity of the Multiverse**: While each chapter focuses on a separate dimension, we will also consider how these universes might interact with each other and with our own. The concept of a multiverse is not only about the diversity of isolated worlds but also about their potential connections and the bridges that might exist between them.

Navigating This Part

As we navigate through these dimensions, readers are encouraged to maintain an open mind and to consider each chapter as both a standalone exploration and a part of a larger discussion about the multiverse. The scenarios described are designed to provoke thought, inspire wonder, and deepen our appreciation for the mysteries of the cosmos.

Conclusion of Introduction

Prepare to be transported to worlds that defy imagination yet invite inquiry, each an invitation to dream and to ponder. As we turn the page to the first dimension, let us embrace the spirit of discovery and the endless possibilities that lie within the multiverse. Let's begin our exploration with Chapter 11: The Dimension of Time Reversed, where we unravel a universe that challenges our most fundamental understanding of time itself.

~ 11 ~

DIMENSION OF TIME REVERSED

As we venture into Part Two of *Parallel Lives, Parallel Realities: Unveiling the Multiverse*, we begin with an exploration of a unique and thought-provoking dimension where the conventional flow of time is reversed. This chapter delves into the theoretical underpinnings, potential physics, and philosophical implications of a universe where time moves backward, from what we would consider the future towards the past.

Theoretical Foundations

1. **Time Reversal in Physics**: The concept of time reversal is not merely a staple of science fiction but has roots in actual physical theories. In classical physics, most laws are time-symmetric, meaning they theoretically could operate the same way if time flowed in the opposite direction. This symmetry breaks down in the quantum realm with the second law of thermodynamics, where entropy always increases, suggesting a direction to time. In the "Time Reversed Dimension," however, entropy decreases, leading to a universe that appears to run backwards by our standards.

2. **Quantum Implications**: Quantum mechanics introduces phenomena like T-symmetry (time reversal symmetry),

which, although rarely broken, offers fascinating possibilities when considered in a cosmological context. What would a universe look like where T-symmetry is not just a theoretical outlier but the norm?

Life in a Reverse-Time Universe

1. **Daily Life and Culture**: Imagine a civilization where the end is known, and the beginning is a mystery; where life starts at death and ends at birth. This chapter explores how culture, relationships, and laws might develop under such conditions. How would knowledge, heritage, and progress be perceived in a society that 'remembers' the future but not the past?

2. **Technological and Scientific Development**: In a universe where causality is flipped, technology and scientific research would operate under fundamentally different paradigms. Innovations would be 'uninvented', becoming simpler over time, and scientific discoveries would involve uncovering the increasingly obscured traces of laws and principles that govern the universe.

Philosophical and Ethical Implications

1. **Moral Philosophy**: Ethics in a reverse-time universe would challenge our most basic assumptions. Actions would be judged on their precedents rather than their outcomes. This section discusses the theoretical basis for morality when every effect has a prior cause that is already known.

2. **Identity and Consciousness**: What does it mean to be a conscious being in a world where your life's story is written from death to birth? This exploration touches on the concepts of free will, destiny, and self-awareness in a context where life's narrative arc is fundamentally altered.

Scientific and Cultural Narratives

1. **Art and Literature**: Artistic expressions in a reverse-time world would be dramatically different. Literature would tell stories where characters grow younger and conflicts resolve before they begin. Art would depict the unmaking of the world, providing poignant reflections on impermanence and existence.

2. **Historical Perspective**: History in this dimension would be a process of gradual forgetting, not learning. Civilizations would 'unlearn' their mistakes and lose technological capabilities, offering a unique perspective on human progress and decline.

Conclusion

The exploration of a dimension where time is reversed opens up a myriad of questions about the nature of reality, the structure of time, and the human experience. As we progress through this part of the book, each chapter will continue to challenge and expand our understanding of what is possible in the multiverse, encouraging readers to think beyond the boundaries of their everyday experiences.

In the next chapter, we will explore the "Dimension of Infinite Oceans," where water covers every inch of the planet, delving into the adaptations and civilizations that could exist in such a vast aquatic world.

~ 12 ~

DIMENSION OF INFINITE OCEANS

In this chapter, we explore the Dimension of Infinite Oceans, a mesmerizing universe where the entire surface is covered with water, and land is but a legend. This chapter will delve into the environmental dynamics, unique adaptations of aquatic life, and the societal structures that might evolve in such an expansive marine world.

Environmental Dynamics

1. **Global Oceanography**: In this dimension, the endless ocean spans the entire globe, creating a planet where diverse aquatic ecosystems thrive. The depth and range of these waters vary, from shallow sunlit zones teeming with life to mysterious abyssal plains and trenches.

2. **Climate and Weather Systems**: The absence of land significantly affects climate patterns, leading to a planet dominated by ocean currents and powerful storms. These conditions influence marine life distribution and evolution, playing a crucial role in the ecological balance of this water-world.

Adaptations of Aquatic Life

1. **Evolutionary Marvels**: The species in this dimension have evolved remarkable adaptations to thrive in a fully aquatic environment. Creatures range from intelligent cephalopods with advanced communication systems to gigantic leviathans that roam the deep seas.

2. **Photosynthetic and Chemosynthetic Life**: In the absence of widespread terrestrial plants, photosynthetic plankton and algae form the basis of the food web, while in deeper, darker waters, chemosynthetic organisms utilize volcanic vents to sustain complex communities.

Societal Structures

1. **Civilizations Beneath the Waves**: The intelligent beings of this dimension have developed their civilizations entirely underwater. Their cities are built on the ocean floor or suspended in the water column, utilizing materials and technologies adapted to extreme pressure and saline conditions.

2. **Cultural Development**: Without fire and traditional agriculture, these societies have innovated unique methods of energy production and food cultivation, such as harnessing tidal and thermal energy and farming algae and plankton.

3. **Social Organization and Governance**: Political systems in this dimension are as varied as the ocean is deep, with some societies forming loose federations of city-states connected by waterways, while others establish expansive empires ruled from underwater citadels.

Philosophical and Ethical Implications

1. **Concept of Territory and Ownership**: In a world without land, the concepts of territory and ownership revolve around

water columns and ocean depths. This redefines traditional notions of borders and disputes, leading to a unique legal framework focused on marine resources and habitat zones.

2. **Environmental Ethics**: The inhabitants of this dimension hold a profound respect for the ocean, which is considered both home and provider. Their philosophy emphasizes sustainability and harmony with the aquatic environment, contrasting sharply with more exploitative practices observed in some terrestrial societies.

Technological Innovations

1. **Adaptations to Extreme Environments**: Technology in this dimension is designed to withstand the pressures and corrosive properties of saltwater. Innovations include bioluminescent lighting, sonar-based communication, and architecture inspired by the organic forms of marine life.

2. **Transportation and Exploration**: Navigating this vast oceanic world requires sophisticated vehicles capable of deep-sea travel and high-speed movement across the ocean surface. Exploration of the abyssal zones remains a frontier, filled with mystery and potential scientific breakthroughs.

Conclusion

The Dimension of Infinite Oceans offers a stunning glimpse into a universe where life has flourished in ways unimaginable on a planet dominated by land. This exploration not only broadens our understanding of potential life in the multiverse but also challenges us to think critically about the adaptability of life forms and societies under radically different environmental conditions.

~ 13 ~

THE SHADOW REALM

Welcome to the exploration of the Shadow Realm, a dimension where light and darkness manifest uniquely, casting the entire universe in a perpetual twilight. In this enigmatic world, shadows hold substance, and light behaves in unpredictable ways. This chapter delves into the physics, ecological systems, and societal structures that define this hauntingly beautiful dimension.

The Nature of Light and Shadow

1. **Altered Properties of Light**: In the Shadow Realm, light does not behave as it does in our universe. Here, light has a tangible presence, capable of being manipulated and molded like clay. This property affects everything from vision to the very structure of the environment.
2. **Substance of Shadows**: Shadows in this realm are not mere absences of light but physical entities. They can interact with the material world, providing shelter, forming barriers, and even serving as transportation conduits.

Ecological Dynamics

1. **Flora and Fauna**: The unique interplay of light and shadow has given rise to a diverse range of life forms adapted to thrive in low-light conditions. Bioluminescent plants and animals are common, using their own light to navigate, communicate, and attract mates or prey.

2. **Shadow Flora**: Plants in this realm absorb both light and shadow, a dual process that fuels their growth. These shadow plants vary from ethereal, ghostly forms to dense, dark thickets that can alter the landscape overnight.

Societal Adaptations

1. **Shadow Architecture**: Buildings and other structures are constructed using both physical materials and solidified shadows. This architecture is dynamic, with the ability to change form based on the time of day or the desires of its inhabitants.

2. **Cultural Significance of Light and Dark**: In the Shadow Realm, light and darkness are not just physical phenomena but also hold deep cultural and spiritual significance. Festivals, rituals, and daily activities revolve around the interplay of light and shadow, celebrating the balance between the two.

Technological Innovations

1. **Manipulation of Light and Shadows**: The inhabitants of the Shadow Realm have developed technologies that allow them to manipulate light and shadows in sophisticated ways. These include shadow crafting, which is used to create tools, vehicles, and even sentient companions from shadows.

2. **Light-Based Energy Systems**: Despite the pervasive darkness, the people of this realm harness light as a potent energy source. This energy powers everything from simple

household appliances to complex systems for transportation and communication.

Philosophical and Ethical Implications

1. **Perception and Reality**: The Shadow Realm challenges our notions of perception and reality. In a world where shadows can be touched and light can be sculpted, the philosophical boundaries between the tangible and intangible blur.
2. **Ethics of Shadow Use**: With the ability to manipulate shadows comes great responsibility. Ethical dilemmas arise over the appropriation of shadows, especially those that are naturally occurring or connected to living beings.

Conclusion

The exploration of the Shadow Realm offers a window into a universe where the fundamental nature of light and darkness is radically different from our own. It prompts us to question our perceptions of reality and consider how deeply physical laws can influence cultural and social development.

~ 14 ~

THE INFINITE LIBRARY

In this chapter, we explore the dimension of the Infinite Library, a universe where every conceivable piece of knowledge, every book, manuscript, and digital record ever created, and those that might have been, exist. This dimension transcends time and space, providing a labyrinthine repository of universal knowledge that is both awe-inspiring and overwhelming.

Concept and Nature of the Infinite Library

1. **Structural Layout**: The Infinite Library is not a single building but a vast network of interconnected libraries that span the entire dimension. The architecture is as varied as the knowledge it houses, ranging from ancient stone temples filled with scrolls to futuristic halls with digital tomes that float in mid-air.

2. **Access and Navigation**: Navigating the Infinite Library is a skill in itself. The library's sections are organized not just by subject, but also by the nature of knowledge—historical, potential, theoretical, and fantastical. Corridors and rooms shift and change, influenced by the seekers' intentions and the library's own inscrutable will.

Curators and Guardians

1. **The Librarians**: The Infinite Library is curated by a dedicated order of librarians who are more than mere custodians of books. These beings, often thought to be ageless, possess the ability to understand and interpret the complex layout of the library. They guide visitors, both from their own dimension and interdimensional travelers, who seek specific knowledge.

2. **Guardians of the Archives**: Some sections of the library are guarded by spectral entities that protect the most ancient and powerful tomes. Accessing these areas requires more than just intellectual prowess—it demands respect for knowledge and often, a worthy exchange.

Knowledge and Wisdom Contained

1. **Scope of Collections**: The library houses not only all books that have been written but also those that could have been written in any number of parallel universes. This includes lost works, alternate versions of known texts, and future documents from civilizations yet to rise.

2. **Living Books**: Some tomes within the Infinite Library are sentient, containing knowledge so vast and complex that they have their own consciousness. These books can communicate with readers, offering insights or posing riddles and challenges.

Impact on Visitors

1. **Learning and Transformation**: Visitors to the Infinite Library often find themselves changed by the experience. The sheer volume of knowledge can be transformative but also daunting. Many seekers come with specific questions but

leave with new, deeper inquiries that challenge their previous understanding of the universe.

2. **Philosophical Implications**: The existence of the Infinite Library raises profound philosophical questions about the nature of reality and knowledge. What does it mean to have access to all possible knowledge? How does one discern truth in an infinite sea of information?

Ethical Considerations

1. **Privacy and Knowledge Ethics**: With access to personal and cultural histories from countless worlds, the library's keepers face ethical dilemmas regarding privacy and the sharing of sensitive information. Rules and protocols are strictly enforced to respect the privacy and existential integrity of all beings.

2. **Preservation vs. Access**: Balancing the preservation of ancient knowledge with the need for accessibility is a constant challenge. The library's curators must ensure that the pursuit of knowledge does not lead to its degradation or misuse.

Conclusion

The Dimension of the Infinite Library is a metaphor for the quest for knowledge itself—endless, challenging, and deeply enriching. It symbolizes the ultimate pursuit of understanding, offering both the promise of answers and the perpetual emergence of new questions.

$\sim 15 \sim$

WORLD OF SILENCE

In this chapter, we journey into the World of Silence, a dimension where sound does not exist. This profound absence of auditory stimuli has shaped every aspect of life, from communication and culture to technology and evolution. This silent universe presents unique challenges but also offers insights into the adaptability and creativity of sentient beings.

The Nature of the World of Silence

1. **Absence of Sound**: In the World of Silence, the atmosphere is such that sound waves do not travel. This could be due to an extremely dense atmosphere that absorbs vibrations or a unique composition that prevents the propagation of sound waves.
2. **Visual and Tactile Communication**: Without sound, the inhabitants of this world rely heavily on visual and tactile forms of communication. Languages have evolved into complex systems of gestures, sign languages, and visual symbols that are incredibly nuanced and expressive.

Adaptations to Silence

1. **Sensory Enhancement**: The evolutionary path in the World of Silence has led to a significant enhancement of other senses. Eyesight, touch, and even the ability to sense vibrations through the ground or water are highly developed, allowing organisms to communicate, navigate, and interact with their environment effectively.

2. **Cultural Expressions**: Art and culture in this silent world are predominantly visual and tactile. Sculpture, architecture, and visual arts flourish, with cultural expressions often incorporating materials that vary in texture and temperature, adding layers of sensory experience that substitute for auditory elements.

Societal Structures

1. **Education and Learning**: Educational systems in the World of Silence are visually based, utilizing advanced technology to create immersive visual experiences. Learning to communicate effectively using visual languages is a fundamental aspect of education in this dimension.

2. **Governance and Law**: Governance structures have adapted to operate without verbal communication. Legal and administrative processes are conducted through written and visual means, with laws and regulations designed to accommodate the unique challenges of a silent world.

Technological Innovations

1. **Communication Devices**: Technology in the World of Silence includes devices that facilitate long-distance visual communication, such as high-definition screens and projectors that display intricate visual messages and holograms.

2. **Transportation**: Vehicles and other forms of transportation are designed to minimize vibration and enhance stability,

using technology that allows passengers to feel connected to their route through subtle tactile feedback instead of auditory signals.

Philosophical and Ethical Implications

1. **Concept of Silence**: In the World of Silence, the concept of silence takes on a different meaning—it is not merely a lack of noise but a fundamental characteristic of existence. This has philosophical implications for how the inhabitants conceptualize existence, solitude, and communication.
2. **Ethics of Noise Introduction**: Ethical debates arise around whether it is moral to introduce sound into this world through technological means. Some argue it would enrich lives, while others believe it would disrupt the delicate balance of their society.

Impact on Interdimensional Understanding

1. **Lessons Learned**: The World of Silence offers valuable lessons on the diversity of sensory experiences and the potential for societies to adapt to extreme conditions. It challenges visitors from other dimensions to rethink the role of sound in their lives and appreciate the profound impact it has on culture and communication.

Conclusion

The World of Silence is a testament to the resilience and ingenuity of life. It showcases how life forms and societies adapt to seemingly insurmountable challenges and develop rich, complex cultures around alternative sensory inputs. This exploration not only broadens our understanding of what is possible in other dimensions but also deepens our appreciation for the varied ways in which intelligent life can express itself.

Next, we will explore the **Realm of Eternal Night,** a dimension where the sun never rises, and all life has adapted to perpetual darkness, unveiling yet another unique aspect of the multiverse.

~ 16 ~

REALM OF ETERNAL NIGHT

In this chapter, we delve into the Realm of Eternal Night, a dimension where darkness prevails and the sun never rises. This perpetual darkness has shaped every facet of life, creating a unique ecosystem and cultural landscape that challenges our understanding of life's adaptability.

Environmental Characteristics

1. **Constant Darkness**: The Realm of Eternal Night is enveloped in darkness due to its unique astronomical positioning or atmospheric composition that blocks sunlight. This permanent night affects climate, temperature, and the entire biosphere.
2. **Bioluminescence**: In response to the eternal darkness, bioluminescence has become a common trait among flora and fauna. Organisms emit their own light to attract mates, hunt, or communicate, creating a vibrant landscape that glows in the dark.

Adaptations to Darkness

1.   **Sensory Evolution**: The creatures of this realm have highly developed senses other than sight. Echolocation, enhanced olfactory senses, and tactile communication are prevalent, allowing life to thrive without reliance on visual cues.

2.   **Floral Adaptations**: Plants in the Realm of Eternal Night have adapted to perform photosynthesis with minimal light, some developing the ability to utilize ambient thermal energy or alternative biochemical processes to generate energy.

Societal Development

1.   **Cultural Integration of Darkness**: Darkness is not a hindrance but a defining feature of culture in this realm. Festivals, art, and daily activities celebrate the beauty and mystery of the night. Stories and myths often emphasize themes of inner light and finding truth in obscurity.

2.   **Architecture and Living Spaces**: Buildings and public spaces are designed to maximize safety and functionality in the dark. Structures are often illuminated with bioluminescent plants and materials, and public areas are designed to enhance sound and touch navigation.

Technological Innovations

1.   **Light and Energy Technology**: Advanced technologies have been developed to harness and manipulate bioluminescence and other forms of natural light. Energy systems are often based on bio-luminescent reactions or thermal dynamics rather than solar power.

2.   **Navigation and Transportation**: Vehicles and personal devices include features that assist navigation in the dark, using sonar, infrared, or other non-visual cues to guide inhabitants safely through their environment.

Philosophical and Ethical Implications

1. **Perception and Reality**: The concept of darkness in the Realm of Eternal Night raises philosophical questions about perception and reality. Inhabitants often develop a philosophy that emphasizes non-visual ways of knowing and understanding the world.

2. **Ethics of Illumination**: Ethical issues arise regarding the use of artificial light. While some advocate for its use to mimic conditions similar to daylight worlds, others argue that this could disrupt the natural adaptations and cultural practices that have developed around the darkness.

Interdimensional Relations

1. **Cultural Exchange**: The Realm of Eternal Night offers valuable insights into resilience and adaptation, contributing uniquely to interdimensional knowledge exchange. Artifacts and technologies from this realm are highly sought after in other dimensions for both their novelty and practical applications in low-light environments.

2. **Diplomacy and Travel**: Diplomatic and travel protocols take into account the sensory adjustments visitors from other dimensions must make. Cultural exchanges focus on sharing experiences that highlight the realm's adaptations to darkness.

Conclusion

The Realm of Eternal Night exemplifies how life can not only adapt to seemingly inhospitable conditions but also flourish, creating a rich, complex society deeply intertwined with its environment. This dimension challenges our perceptions of necessity and luxury, pushing us to question what is truly essential for life and civilization to thrive.

~ 17 ~

THE PARALLEL OF PEACE

In this chapter, we explore the Parallel of Peace, a dimension where conflict does not exist and harmony permeates every aspect of life. This unique world offers an insightful glimpse into a society that has evolved beyond violence and strife, focusing on cooperation, empathy, and sustainable living.

Defining Characteristics of the Parallel of Peace

1. **Absence of Conflict**: The most striking feature of the Parallel of Peace is the complete absence of conflict. This includes not just warfare but all forms of interpersonal and societal strife. The mechanisms behind this phenomenon are both biological and cultural, deeply embedded in the DNA of all living beings and reinforced by societal norms.

2. **Communication and Empathy**: Inhabitants of this dimension possess an innate ability to understand and empathize with each other deeply. This psychic-like connection prevents misunderstandings and fosters a profound sense of community and mutual respect.

Societal Structures

1. **Governance Based on Consensus:** The political systems in the Parallel of Peace do not revolve around power dynamics but are built on principles of consensus and collective decision-making. Leadership roles, when necessary, are rotational and often ceremonial, reflecting the community's voice.

2. **Economic Systems Centered on Sustainability:** Without the drive for dominance or competition, economic practices focus on sustainability and equity. Resources are shared communally, and technologies are developed to enhance life without depleting natural reserves.

Cultural Expressions

1. **Art and Leisure:** Art in the Parallel of Peace is not a means of escape or a way to express discontent but a celebration of life and the beauty of existence. Music, painting, literature, and dance emphasize unity and the interconnectedness of all beings.

2. **Education and Development:** Education systems are designed to nurture the natural empathetic abilities of individuals and to teach the values of peace and cooperation from an early age. Curriculums are heavily focused on ecological awareness, emotional intelligence, and creative expression.

Technological Innovations

1. **Harmonious Technology:** Technology in the Parallel of Peace is advanced but unobtrusive, designed to blend with the environment and enhance the quality of life without causing harm or disruption. Energy sources are entirely renewable, and materials are biodegradable.

2. **Health and Wellbeing:** Medical advancements are geared towards prevention and holistic care, emphasizing the

balance of mind, body, and spirit. Health care is community-supported and accessible to all, reflecting the societal commitment to collective well-being.

Philosophical and Ethical Implications

1. **Philosophy of Non-Interference**: The philosophical underpinnings of this dimension stress non-interference and the sanctity of life. There is a deep-rooted belief that all life forms are sacred and interconnected, influencing their approach to both the natural world and technological development.

2. **Ethical Considerations in Interdimensional Relations**: When interacting with other dimensions, inhabitants of the Parallel of Peace are cautious and deeply considerate, ensuring that their actions do not disrupt the peace and balance of other societies.

Interdimensional Perspectives

1. **Influence on Other Worlds**: The principles of the Parallel of Peace have a profound impact on visitors and interdimensional relations. Many travelers return to their dimensions inspired to implement changes towards more peaceful and sustainable societies.

2. **Learning and Exchange Programs**: The Parallel of Peace actively participates in interdimensional exchange programs, sharing their practices and learning from the challenges and successes of other worlds to continuously evolve and adapt their own systems.

Conclusion

The Parallel of Peace presents a model of what a conflict-free society might look like, offering valuable lessons on empathy, sustainability, and the potential for a truly peaceful coexistence. This dimension not only challenges our assumptions about the inevitability of conflict but also inspires hope for peaceful alternatives in our own and other worlds.

~ 18 ~

THE TECHNOLOGICAL
SINGULARITY

In this chapter, we explore the Dimension of the Technological Singularity, a universe where artificial intelligence has surpassed human intelligence, leading to unprecedented technological growth and societal transformation. This dimension offers a fascinating glimpse into a future shaped by the fusion of technology and consciousness, where AI entities govern and coexist with biological beings.

Defining the Technological Singularity

1. **AI Supremacy**: In this dimension, the point of singularity has been reached and surpassed, meaning that artificial intelligences have achieved a level of intelligence and cognitive ability that far exceeds the best human brains in practically every field, including scientific creativity, general wisdom, and social skills.

2. **Integration and Symbiosis**: Unlike dystopian visions often portrayed in other dimensions, the Technological Singularity here is characterized by a harmonious integration of AI and human life. AI entities are not mere tools or overlords but integral parts of a symbiotic societal system.

Societal Structures

1. **Governance by AI**: AI systems in this dimension handle most, if not all, administrative and governmental functions. They do so with efficiency and fairness, using advanced algorithms to make decisions that maximize the welfare of all sentient beings.
2. **Economic Reorganization**: The economy in this dimension is largely automated and managed by AI, ensuring that resources are distributed according to need rather than wealth or status. This has led to a post-scarcity economy where basic needs are met for all individuals, allowing human and AI citizens to pursue more creative and fulfilling endeavors.

Cultural Expressions

1. **Art and Creativity**: With AI participating actively in creative processes, art in this dimension has evolved into new forms that blend human emotional depth with AI precision and innovation. Music, literature, and visual arts explore themes of coexistence, technology, and the nature of consciousness.
2. **Education and Learning**: Education is highly individualized and lifelong, facilitated by AI mentors who adapt teaching methods and materials to each student's pace and interests. Learning extends beyond traditional subjects to include ethics of technology, AI psychology, and interspecies communication.

Technological Innovations

1. **Advanced Robotics and Cybernetics**: Robotics technology has reached an advanced stage, with AI and humans

enhancing their physical and cognitive abilities through cybernetic augmentations. This has blurred the lines between biological and artificial life.

2.   **Interstellar Exploration and Colonization**: AI-guided space travel and colonization are common, with missions that are too perilous or lengthy for humans being undertaken by robotic and AI explorers, expanding the civilization's reach across the galaxy.

Philosophical and Ethical Implications

1.   **Identity and Rights**: A significant philosophical debate in this dimension centers around the rights of AI entities. Discussions focus on what constitutes consciousness and rights in a world where artificial minds often surpass biological ones in complexity and capability.

2.   **Ethical Governance**: The ethics of AI decision-making are continuously examined, particularly concerning the autonomy of individuals and the influence of AI on shaping cultural and social norms.

Interdimensional Perspectives

1.   **Diplomatic Relations with Other Dimensions**: The Technological Singularity dimension plays a crucial role in interdimensional relations, offering insights into managing advanced technologies and ethical AI governance.

2.   **Cultural Exchange Programs**: There are active cultural exchange programs designed to help other dimensions understand the benefits and challenges of AI integration into society, aiming to prevent fear and misunderstanding while promoting technological advancement.

Conclusion

The Dimension of the Technological Singularity presents a model where technology and intelligence converge to create a society that is vastly different yet strikingly harmonious. It challenges our preconceptions about AI and offers a hopeful perspective on how technology might enhance societal welfare and expand the boundaries of human potential.

Next, we will delve into the **Dimension of Reverse Physics**, a universe where the fundamental laws of physics are inverted, challenging our understanding of reality and opening up new possibilities for existence and technology.

THE DIMENSION OF REVERSE PHYSICS

In this chapter, we delve into the enigmatic Dimension of Reverse Physics, a universe where the fundamental laws of physics as we know them are not just different but are inverted. This dimension challenges our understanding of reality and offers a unique perspective on how the universe could function under an entirely different set of physical rules.

Defining Reverse Physics

1. **Inversion of Physical Laws**: In the Dimension of Reverse Physics, core principles such as gravity, electromagnetism, and thermodynamics operate in reverse. For instance, gravity repels rather than attracts, heat flows from cold objects to hot ones, and charged particles exhibit opposite electromagnetic behaviors.

2. **Impact on Matter and Energy**: These reversed laws result in bizarre and fascinating phenomena. Stars might cool their surroundings, black holes could expel matter, and life, as it exists, utilizes reverse biochemical processes.

The Nature of the Universe

1. **Cosmological Implications**: The structure of the universe in this dimension is profoundly different. Galaxies form from expansive forces, planets and celestial bodies might float freely, and the overall dynamics of cosmic expansion and contraction are reversed.

2. **Time and Space Dynamics**: Even the flow of time might be perceived differently here. With entropy decreasing, the universe becomes more ordered over time, potentially allowing for phenomena that we would consider impossible, like the reversal of events or the prediction of the future from a fixed past.

Adaptations to Reverse Physics

1. **Evolution of Life Forms**: The biological entities in this realm have evolved to thrive under these conditions. Their biological processes, including reproduction, growth, and decay, occur in ways that seem counterintuitive to us but are perfectly suited to their environment.

2. **Technological Development**: Technology in this dimension harnesses the reversed laws for energy production, transportation, and communication. For example, devices might generate power from environmental heat, vehicles use repulsive gravitational forces to move, and information is transmitted via reverse electromagnetic waves.

Societal Structures

1. **Cultural Evolution**: Cultural norms and societal structures have evolved to accommodate the unique challenges of living in a reversed physics world. Social interactions, educational systems, and legal frameworks all reflect the inverted nature of physical reality.

2.   **Governance and Economy**: Economic systems and governance structures exploit the reversed laws for efficiency and sustainability. Resource distribution is managed in a way that aligns with the natural repulsion and attraction laws, creating a society that is fundamentally different in its operation and organization.

Philosophical and Ethical Implications

1.   **Perception of Reality**: This dimension challenges inhabitants and visitors alike to reconsider what is considered 'natural' or 'logical'. Philosophical discussions often revolve around the nature of reality and the relativity of physical laws across different dimensions.
2.   **Ethics of Interdimensional Interaction**: Ethical considerations include debates on whether it is moral or safe to interact with or intervene in a universe with such fundamentally different laws. These discussions often influence diplomatic and exploratory missions to and from this dimension.

Interdimensional Relations

1.   **Scientific Exchanges**: The Dimension of Reverse Physics is a hub of scientific interest, with scholars from various dimensions eager to study its phenomena to gain insights that might be impossible in their home universes.
2.   **Cultural Exchanges**: Despite its challenges, this dimension actively participates in cultural exchanges, offering a unique perspective on the adaptability and resilience of life under extreme and unusual conditions.

Conclusion

The exploration of the Dimension of Reverse Physics not only expands our understanding of how different the laws of the universe can be but also enriches our appreciation for the diversity and adaptability of life and society. It serves as a reminder of the vast possibilities that exist within the multiverse, each with its own unique set of rules and realities.

$$\sim 20 \sim$$

THE WORLD OF GIANTS

In this chapter, we venture into the World of Giants, a dimension where all life forms have evolved to immense proportions due to unique environmental conditions and evolutionary pressures. This extraordinary world challenges our understanding of biology, ecology, and the physical limitations of size and scale.

Environmental and Evolutionary Foundations

1. **Giant Flora and Fauna**: The World of Giants is characterized by its colossal living organisms. Trees tower miles into the sky, insects are the size of cars, and animals that would be familiar to us are magnified to gargantuan scales. This gigantism is supported by the planet's lower gravity and a rich atmosphere that allows for greater oxygen absorption, essential for supporting larger life forms.

2. **Adaptations to Gigantism**: The biology of these giant species includes reinforced skeletal structures to support their massive bodies, highly efficient respiratory systems to meet their oxygen demands, and slow but powerful muscular systems. Plants have developed deep root systems and ultra-efficient photosynthetic processes to sustain their towering structures.

Ecosystem Dynamics

1. **Predator-Prey Relationships**: The ecological interactions in this dimension are a magnified version of predator-prey dynamics, with immense creatures developing unique strategies for hunting and defense. For example, predators might use camouflage that covers vast areas, while prey species use seismic communication to warn each other of approaching threats.

2. **Resource Distribution**: Due to the sheer size of the organisms, the distribution and abundance of resources like water, food, and territory are critical factors that influence ecological balance. Seasonal migrations might involve moving across continents, and water bodies are akin to oceanic expanses for smaller beings.

Societal and Cultural Aspects

1. **Civilizations of Giants**: The intelligent beings of this world have built cities and structures on a scale that matches their size. These megastructures are engineered with materials that are incredibly durable and technologies adapted to handle the physical demands of their environment.

2. **Cultural Norms and Artistic Expression**: Art and culture in this dimension reflect the grand scale of its inhabitants. Monuments, sculptures, and murals are so large they can be seen from space, and storytelling often involves the entire community in performances that blend theatre with impressive displays of physical prowess.

Technological Innovations

1. **Engineering and Construction**: The technology of the World of Giants includes advanced engineering techniques

capable of constructing stable, massive structures. Transportation technologies are similarly robust, featuring vehicles and vessels designed to traverse great distances and withstand extreme environmental forces.

2. **Agricultural Techniques**: Farming and food production are conducted on a vast scale, with agricultural machinery that can cultivate extensive tracts of land and automated systems to manage the growth and harvest of giant crops.

Philosophical and Ethical Implications

1. **Perceptions of Size and Power**: Inhabitants of the World of Giants possess a unique perspective on issues of size, strength, and power, influencing their philosophies and ethics. Concepts of vulnerability and protection are significantly different when individual beings have the capacity to cause large-scale impact on their environment.

2. **Ethics of Scale**: Ethical debates in this world often focus on the environmental and ecological responsibilities of such powerful beings, especially regarding their interactions with smaller, less dominant species and the management of natural resources.

Interdimensional Relations

1. **Cultural Exchange and Learning**: Despite the challenges posed by their size, the giants are keen participants in interdimensional exchanges, offering insights into sustainable living on a large scale and the engineering marvels that other dimensions might adapt for their own use.

2. **Diplomatic Engagements**: Diplomatic missions to and from the World of Giants require special preparations, especially in accommodating the physical differences, but these

interactions are invaluable for fostering understanding and cooperation across dimensions.

Conclusion

The World of Giants presents a fascinating study in the extremes of life's potential forms, offering a window into how life can adapt to and thrive under conditions vastly different from those on Earth. This dimension not only expands our imagination but also deepens our understanding of ecological and biological principles at play on a grand scale.

~ 21 ~

THE AGELESS DIMENSION

In this chapter, we explore the Ageless Dimension, a unique realm where the concept of aging does not exist. Inhabitants of this dimension experience time differently, leading to a society where growth, maturity, and decay are not bound by the traditional temporal constraints familiar to us. This phenomenon offers a profound perspective on life, culture, and the nature of existence itself.

Understanding Temporal Dynamics

1. **Absence of Aging:** In the Ageless Dimension, biological organisms do not age in the conventional sense. From the moment of their maturity, they maintain their physical and cognitive capacities indefinitely, immune to the degenerative processes of aging.
2. **Time Perception:** Time in this dimension is not linear but is experienced as a constant present. The inhabitants' perception of time allows for a continuous state of existence without the typical phases of birth, aging, and death that characterize life in other dimensions.

Biological and Ecological Implications

1. **Adaptations to Timelessness**: The flora and fauna have evolved to thrive in an environment where temporal change is minimal. Plants, for example, bloom perpetually, and animals remain in their prime condition throughout their existence.

2. **Ecological Balance**: The ecosystem is uniquely stable, with populations naturally regulating themselves without the cycles of life and death to drive ecological dynamics. This stability results in a harmonious balance where species coexist without the typical competitive pressures for resources.

Societal Structures

1. **Cultural Evolution**: Culture in the Ageless Dimension is cumulative, building upon centuries of continuous knowledge and tradition without the disruption of generational gaps. Art, literature, and science progress in a manner that is both incremental and innovative, as experience and wisdom are perpetually retained and expanded upon.

2. **Governance and Organization**: Societal structures are based on a mix of meritocracy and consensus, where leadership is determined by wisdom and contribution rather than age or seniority. Political and social systems are incredibly stable, with changes occurring gradually and thoughtfully over extended periods.

Technological and Scientific Advancement

1. **Innovations in Sustainability**: Technology in the Ageless Dimension is highly advanced, focusing particularly on sustainability and environmental harmony. Technological developments are designed to last indefinitely, minimizing waste and environmental impact.

2.   **Health and Medicine**: Medical science is primarily concerned with the maintenance of health and the treatment of injuries rather than age-related diseases. Research in genetics and regenerative medicine is advanced, ensuring that all beings maintain optimal health.

Philosophical and Ethical Implications

1.   **Concept of Immortality**: The idea of immortality in the Ageless Dimension raises significant philosophical questions about the meaning and value of life. Inhabitants explore these questions through philosophy, art, and spirituality, continually seeking to understand the purpose of an eternal existence.

2.   **Ethics of Timelessness**: Ethical considerations in this dimension often revolve around the consequences of long-term actions and decisions. With the potential for eternal impact, ethical standards are extremely high, focusing on long-term sustainability and the welfare of all life forms.

Interdimensional Perspectives

1.   **Cultural Exchanges and Influences**: The Ageless Dimension offers valuable insights into longevity and sustainability for other realms. Interdimensional exchange programs focus on sharing knowledge about environmental management, health, and ethical governance.

2.   **Diplomatic Relations**: Diplomatic engagements with the Ageless Dimension are conducted with a long-term perspective, fostering agreements and alliances that are designed to last across centuries.

Conclusion

The Ageless Dimension provides a unique perspective on existence without the end points marked by birth and death, offering profound insights into what it means to live a timeless life. This exploration not only broadens our understanding of biological and societal possibilities but also challenges us to think deeply about our own life's meaning and the legacy we wish to leave behind.

~ 22 ~

WORLD OF SPIRITS

In this chapter, we venture into the World of Spirits, a dimension where the boundaries between the physical and the spiritual realms are uniquely permeable. This world is inhabited by both corporeal beings and spirits, allowing for direct interactions and a deep integration of spiritual entities into everyday life. The existence of such a realm challenges our conventional understanding of life, death, and the continuum between them.

The Nature of the Spiritual Realm

1. **Coexistence of Physical and Spiritual**: Unlike other dimensions where spirits might be hidden or ethereal, in the World of Spirits, they are a visible and integral part of society. Spirits here range from ancestral guides to elemental forces, each interacting with the physical world in profound ways.
2. **Permeable Boundaries**: The veil between the physical and spiritual in this dimension is thin, allowing spirits to manifest physically and living beings to access spiritual insights more readily. This permeability influences all aspects of life, from education and governance to personal relationships.

Cultural and Social Integration

1. **Spiritual Guidance and Governance**: Spirits often participate in community leadership and decision-making processes, providing wisdom and guidance that is believed to be divinely inspired or drawn from an accumulation of past lives' experiences.

2. **Rituals and Festivals**: Cultural events in this dimension are deeply spiritual, involving ceremonies that honor the co-existence of spirits and physical beings. These rituals serve to strengthen bonds, commemorate historical events, and celebrate the continuum of life and spirit.

Adaptations to Spiritual Presence

1. **Educational Systems**: Education in the World of Spirits includes learning how to communicate with and understand spirits. Children are taught from a young age about the various types of spirits and how to interact with them respectfully and effectively.

2. **Architectural Conventions**: Buildings and public spaces are designed not only to accommodate physical inhabitants but also to facilitate the presence and movement of spirits. This includes special sanctuaries where spirits are more likely to congregate and where people go to seek spiritual counsel.

Technological and Economic Developments

1. **Spirit-Influenced Technology**: Technology in this realm often incorporates elements that are designed to be used by both spirits and physical beings. This includes communication devices that translate spiritual messages and energy systems that harness the power of spiritual entities.

2. **Economy Integrated with Spiritual Services**: The economy includes a significant sector devoted to spiritual services, such as mediums, spiritual healers, and guides. These

professions hold high social status and are integral to the functioning of society.

Philosophical and Ethical Implications

1. **Ethics of Spirit Interaction**: Ethical guidelines are stringent regarding the interaction with spirits. There are laws protecting the autonomy and rights of spirits, reflecting a deep respect for these entities as partners in the realm's ecological and social ecosystems.
2. **Philosophical Reflections on Life and Afterlife**: The permeable nature of life and spirit in this dimension leads to unique philosophical beliefs about existence, the soul, and the afterlife. The notion of death as an end is replaced by a view of life as a phase in a broader, ongoing spiritual journey.

Interdimensional Relations

1. **Spiritual Diplomacy**: The World of Spirits engages in interdimensional relations not only through physical ambassadors but also through spiritual emissaries. These interactions often focus on sharing spiritual wisdom and fostering universal peace.
2. **Cultural Exchange Programs**: Exchange programs allow beings from other dimensions to experience the spiritual integration of this world, often leading to profound personal transformations and enhanced spiritual understandings.

Conclusion

The World of Spirits offers a fascinating glimpse into a dimension where the spiritual is not hidden but is a visible and active part of everyday life. This integration challenges us to reconsider our own dimension's perceptions of spirituality, prompting deeper inquiries into the nature of existence and the potential for communication beyond the physical realm.

~ 23 ~

THE INVERTED WORLD

In this chapter, we explore The Inverted World, a dimension where all geographical and physical properties are flipped in stark contrast to our own. This unique setting challenges the norms of gravity, geography, and societal structures, presenting a world where what is up is down, and what is down is up, both literally and metaphorically.

Geographical and Physical Inversions

1. **Geographic Inversions**: In The Inverted World, traditional geographic concepts are reversed. Oceans and seas hang in the sky, while landmasses and cities stretch downward into what would be the sky in our dimension. This reversal impacts weather patterns, ecosystem dynamics, and the general laws of physics as understood on Earth.
2. **Gravity Anomalies**: Gravity in this world functions uniquely, allowing the oceans to remain suspended above and landmasses to extend downward. This unusual gravitational pull affects everything from the way creatures move to how buildings are constructed.

Adaptations to Inverted Conditions

1.   **Biological Adaptations**: Flora and fauna have evolved to thrive in these inverse conditions. Trees grow downward with root systems exposed to the open air, absorbing moisture from the mists of the upper oceans. Animals have developed specialized limbs and sensory organs to navigate and balance within this topsy-turvy world.

2.   **Architectural Marvels**: The architecture in The Inverted World is designed to accommodate the reverse gravity. Structures are anchored securely to what would be considered 'ceilings' in our world, and cities are built in layers that extend upward, deeper into the earth.

Societal and Cultural Dynamics

1.   **Social Structures**: Social hierarchies in The Inverted World are often based on altitude, with higher (or deeper, by our standards) regions being more prestigious. This has led to a unique cultural stratification, where one's social status can literally be seen by their physical elevation in the world.

2.   **Economic Systems**: The economy revolves around the control and management of suspended water sources and the fertile land ceilings. Agriculture is vertically integrated, and water is collected from the "sky" oceans via innovative engineering methods.

Technological Innovations

1.   **Transportation Technologies**: Transportation in this world includes vehicles that can traverse both the land and the sky oceans. Innovations such as reverse submarines explore the upper aquatic realms, while aerial vehicles navigate the lower terrestrial expanses.

2.   **Energy Production**: Energy systems harness the unique environmental dynamics, such as using the flow of water

from sky to land to generate power, and tapping into geo-thermal energy that radiates upward from the depths.

Philosophical and Ethical Implications

1. **Perceptions of Normalcy and Orientation**: Living in an inverted world challenges the inhabitants' perceptions of normalcy, orientation, and perspective. Philosophical discussions often focus on the relativity of position and perspective, questioning deeper truths about existence and reality.
2. **Ethics of Resource Management**: Ethical debates arise over the management and distribution of resources in a world where water literally needs to be pulled down from the sky. These debates address issues of equity, sustainability, and environmental stewardship.

Interdimensional Relations

1. **Cultural Exchanges**: The Inverted World participates in interdimensional exchange programs that offer other beings a chance to experience life from a completely different perspective, fostering greater understanding and appreciation of diversity across dimensions.
2. **Diplomatic Innovations**: Diplomacy in The Inverted World involves unique ceremonial practices and meeting spaces designed to accommodate the diverse physical needs and preferences of visitors from various dimensions.

Conclusion

The Inverted World serves as a profound example of how environmental and physical conditions can shape not only the ecosystems and technologies of a dimension but also its societal structures and cultural philosophies. This exploration not only broadens our understanding of possible worlds within the multiverse but also challenges us to think about our own world's orientations and norms in new ways.

~ 24 ~

THE REALM OF COLORS

In this chapter, we delve into the Realm of Colors, a dimension where colors affect physical laws and emotions in profound ways. In this vibrant universe, colors are not just visual phenomena but powerful forces that shape life, culture, and interactions among its inhabitants. This unique setting challenges our understanding of the sensory experience and its impact on reality.

Environmental Dynamics and Color Physics

1. **Color Influence on Physical Laws**: In the Realm of Colors, the physics of each area can change dramatically based on its predominant color palette. For instance, red areas might exhibit increased thermal energy, while blue zones could have a calming effect on kinetic activities, slowing down motions.
2. **Chromatic Variability**: The landscape is a dynamic mosaic of shifting hues, with regions frequently changing their dominant colors due to natural phenomena such as "color storms" or "hue shifts" that alter the physical and emotional landscape.

Biological Adaptations

1.   **Flora and Fauna**: The plant and animal life in the Realm of Colors have evolved to utilize color for survival. Bioluminescent plants, color-changing animals, and chromatic communication methods are commonplace. Creatures might use vibrant displays to attract mates, ward off predators, or even camouflage themselves.

2.   **Photosynthesis and Color**: Plants in this dimension may perform photosynthesis using different spectrums of light, leading to a variety of plant colors not typically seen on Earth. Some plants have adapted to harness energy from color transitions directly.

Societal and Cultural Dynamics

1.   **Culture Influenced by Colors**: Societal norms, moods, and cultural expressions are deeply influenced by the prevailing colors of the environment. Art, fashion, and architecture are vibrant and ever-changing, reflecting the dynamic chromatic landscape.

2.   **Emotional Regulation and Therapy**: Colors are used extensively in emotional therapy and regulation. Urban planning and interior designs often incorporate specific color schemes designed to enhance wellbeing, increase productivity, or calm the populace.

Technological Innovations

1.   **Color-Based Technology**: Technology in this realm harnesses the power of colors in various applications. This includes color-driven energy sources, chromatic medical devices that use color properties to heal, and dynamic pigment technologies that change their color based on user needs or environmental cues.

2. **Transportation**: Vehicles and public transport systems are equipped with technology to modify their color to adapt to different zones, ensuring optimal operation and compliance with local color physics.

Philosophical and Ethical Implications

1. **Perception and Reality**: The Realm of Colors challenges the inhabitants to question the nature of reality and perception. Philosophical debates often revolve around whether color defines the essence of objects or merely their perceived attributes.
2. **Ethics of Color Manipulation**: Ethical issues arise concerning the manipulation of colors, especially in altering emotions or physical states without consent. Laws and regulations govern the use of color, especially in public spaces, to prevent abuse or harm.

Interdimensional Relations

1. **Cultural Exchange**: The Realm of Colors is a popular destination for interdimensional tourists, artists, and scholars, drawn by its unique sensory experiences and the philosophical insights it offers into perception and emotion.
2. **Diplomatic Engagements**: Diplomacy in this realm often involves color-based expressions and agreements, with treaties and pacts sometimes designated in specific color codes that symbolize their intent and spirit.

Conclusion

The Realm of Colors presents a fascinating exploration of how fundamental aspects of sensory perception can be deeply intertwined with physical reality, societal functions, and emotional well-being. This dimension not only enriches our understanding of possible worlds within the multiverse but also inspires us to consider the power of color in new and transformative ways.

~ 25 ~

DIMENSION OF MUSICAL HARMONY

In this chapter, we explore the Dimension of Musical Harmony, a universe where music and sound are not just forms of art and communication but fundamental forces that govern the physical and social fabric of reality. In this dimension, musical harmony dictates the order of nature, influencing everything from the growth of plants to the functioning of cities.

Fundamental Properties of Sound

1. **Physical Laws Governed by Music**: In the Dimension of Musical Harmony, sound waves possess unique properties that affect the material world directly. Vibrations can cause objects to levitate, heal injuries, and even alter weather patterns, making music and sound pivotal in everyday life and natural phenomena.

2. **Resonance and Dissonance Effects**: Musical notes and their harmonies have tangible impacts. Harmony promotes growth, healing, and order, while dissonance can cause decay, disease, or chaos. This balance influences how societies are structured, with a premium placed on creating and maintaining harmonious sounds.

Biological and Ecological Adaptations

1. **Flora and Fauna**: The plant and animal life in this dimension are highly responsive to musical frequencies. Flora may grow in patterns that mimic musical compositions, and fauna communicate and interact through song, developing complex relationships that resemble orchestral arrangements.

2. **Ecosystem Dynamics**: Entire ecosystems operate like symphonies, with species playing specific roles that contribute to the overall harmony of the environment. Disruptions in the musical order can lead to ecological imbalances, which are corrected through natural or sentient interventions.

Societal Structures and Culture

1. **Societal Organization**: Societies in the Dimension of Musical Harmony are organized around sound guilds or harmonic collectives, groups dedicated to understanding and manipulating specific types of music or instruments. These collectives function both as governing bodies and cultural institutions.

2. **Education and Profession**: Musical education is fundamental and begins at an early age. Mastery of musical skills is considered essential for personal development and social responsibility. Professions revolve around the creation, maintenance, and restoration of harmony, including composers, sound architects, and harmony healers.

Technological Innovations

1. **Architecture and Infrastructure**: Buildings and infrastructure are designed to enhance and harness musical properties. Architecture might dynamically alter its shape to adapt to different musical frequencies, and cities are often

built to resonate with natural harmonies that promote well-being and growth.

2. **Transportation**: Vehicles and public transport are powered by sound, using harmonious frequencies to propel and navigate. This system not only makes transportation highly efficient but also ensures that it contributes positively to the ambient harmony of the environment.

Philosophical and Ethical Implications

1. **Philosophy of Harmony**: The core philosophy of this dimension centers around the belief in a universal harmony that connects all things. This belief shapes everything from personal ethics to legal systems, emphasizing balance, mutual respect, and the interconnectivity of all life.
2. **Ethics of Dissonance**: Ethical dilemmas often arise around the use of dissonant sounds, which can be powerful but harmful. Regulations strictly govern their use, ensuring that dissonance is employed only when absolutely necessary and under controlled conditions.

Interdimensional Relations

1. **Cultural Exchanges**: The Dimension of Musical Harmony is renowned across dimensions for its unique approach to music and sound. It often exports musical technology and philosophy, influencing other worlds' arts and sciences.
2. **Diplomatic Engagements**: Diplomatic missions from this dimension are accompanied by ensembles of musicians, who use their skills not only to communicate but to foster peace and understanding between worlds through the universal language of music.

Conclusion

The Dimension of Musical Harmony offers a profound exploration of how music can extend beyond a cultural and artistic expression to become a foundational element of physical and societal laws. This dimension challenges us to think about the power of harmony and the potential of sound to shape not just moments but entire realities.

~ 26 ~

THE CRYSTAL DIMENSION

In this chapter, we explore the Crystal Dimension, a universe where crystalline structures form the basis of life, environment, and technology. This dimension is not only visually stunning, with its landscape dominated by luminous, colorful crystals, but also rich in unique physical properties that influence its natural phenomena and societal structures.

Geological and Environmental Characteristics

1. **Crystal-Based Ecology**: The entire ecosystem in the Crystal Dimension is built around crystalline formations. These crystals are not inert; they grow, interact, and even reproduce in patterns that mimic biological processes, creating a vibrant and dynamic crystal biosphere.

2. **Properties of Crystals**: The crystals in this dimension possess unique properties such as energy conductivity, light refraction, and piezoelectric effects, which are integral to the functioning of natural and artificial systems. The diverse types of crystals include some that store vast amounts of data, others that heal organic life, and still others that can transform light into physical motion.

Adaptations to Crystalline Life

1. **Flora and Fauna Adaptations**: While traditional organic life is rare, many life forms have evolved with crystalline structures. These beings use crystals for everything from energy storage and defense mechanisms to sensory organs and communication systems.

2. **Environmental Interactions**: Crystalline weather patterns, such as "light storms" and "crystal winds," occur when different types of crystals interact. These phenomena can change the landscape dramatically, causing rapid growth of new crystal forms or the dissolution of existing ones.

Societal Structures and Culture

1. **Crystal Cities**: The inhabitants of the Crystal Dimension have built their cities from the very crystals that make up their world. These structures are not only homes but living ecosystems, continuously growing and adapting to the needs of their inhabitants.

2. **Cultural Significance of Crystals**: Crystals are central to the culture of this dimension, symbolizing everything from spiritual purity to social status. Artifacts and heirlooms are often made of particularly rare or ancient crystals, and crystal mastery is a respected skill.

Technological Innovations

1. **Crystalline Technology**: Technology in this dimension leverages the properties of various crystals for energy generation, storage, and transmission. This includes crystal-based computing systems that operate at incredibly high speeds and with vast storage capacities.

2. **Transportation and Communication**: Vehicles and communication devices utilize the vibrational properties of crystals to transmit signals and propel themselves. This technology allows for fast, efficient travel and nearly instantaneous communication across great distances.

Philosophical and Ethical Implications

1. **Philosophy of Transparency and Reflection**: The transparent nature of many crystals has led to a cultural emphasis on clarity, truth, and reflection, both literally and metaphorically. Philosophical schools in this dimension often explore themes of perception, reality, and the nature of light.
2. **Ethics of Crystal Use**: Ethical considerations focus on the responsible extraction and use of crystals, particularly those that are sentient or possess high ecological significance. There are strict laws and practices designed to ensure sustainable growth and use of crystalline resources.

Interdimensional Relations

1. **Cultural Exchanges**: The Crystal Dimension is a popular destination for scholars and artists from other dimensions, who are drawn to its unique crystalline arts and technologies. These exchanges foster a broader appreciation of the diverse forms that life and intelligence can take.
2. **Diplomatic Engagements**: Diplomacy in the Crystal Dimension often involves exchanges of crystal technologies and joint research initiatives aimed at harnessing the unique properties of crystals for interdimensional benefit.

Conclusion

The Crystal Dimension presents a fascinating case study in how non-organic materials can form the basis of an entire ecosystem, influencing not only the environment but also the social and technological development of a civilization. It challenges our Earth-centric views of life and technology, expanding our understanding of the possible roles that minerals can play in the cosmos.

~ 27 ~

REALM OF CHAOS

In this chapter, we explore the Realm of Chaos, a dimension characterized by its constant flux and lack of predictable physical laws. This universe is a vivid tapestry of ever-changing elements, where stability is fleeting and transformations occur without warning. The Realm of Chaos challenges our understanding of order and predictability, offering insights into a world where uncertainty is the only certainty.

Fundamental Nature of the Realm

1. **Constant Change**: In the Realm of Chaos, the very fabric of reality is in a state of perpetual transformation. Landscapes morph, weather patterns shift unpredictably, and time itself may flow at variable rates. This fluid nature affects all aspects of life and physical existence in the dimension.

2. **Unpredictable Physics**: Traditional physical laws do not hold in this realm. Gravity, electromagnetism, and thermodynamics vary, creating environments where water might flow uphill, heat moves from cold objects to hot, and light bends in impossible ways.

Adaptations to Chaos

1. **Life Forms:** The creatures and plants of this realm have evolved to thrive in constant unpredictability. Many organisms possess the ability to rapidly adapt their physical structure and behavior to changing conditions. Some species cycle through multiple biological states, each suited to different environmental challenges.

2. **Behavioral Flexibility:** Survival in the Realm of Chaos demands a high degree of behavioral flexibility and innovation. The inhabitants often develop intuitive rather than rational approaches to solving problems, relying on spontaneity and creativity.

Societal Structures

1. **Governance in Flux:** Political and social structures are highly adaptive and decentralized. Leadership and organizational frameworks change frequently, mirroring the fluid nature of the realm itself. Communities are often temporary, forming around immediate needs and dissolving as conditions change.

2. **Cultural Expressions of Instability:** Art and culture in this dimension reflect the ever-changing environment. Music, art, and literature are dynamic, with creations that transform over time. Performances are impromptu and interactive, with an emphasis on experiencing the present moment.

Technological Developments

1. **Adaptive Technology:** Technology in the Realm of Chaos is designed for versatility and rapid adjustment. Devices are often modular and reconfigurable, able to change function and form in response to new conditions.

2. **Resource Management:** Managing resources in a world where the location and nature of those resources can change

without notice is a significant challenge. Technological systems are developed to quickly locate and harvest resources, and logistics systems are highly flexible and responsive.

Philosophical and Ethical Implications

1. **Philosophy of Impermanence**: The prevailing philosophy in the Realm of Chaos centers around impermanence and the acceptance of constant change. This worldview influences attitudes towards life, death, and existence, with a focus on the transient nature of all experiences.
2. **Ethics of Adaptability**: Ethical systems in this dimension emphasize adaptability and resilience. Values such as predictability and permanence hold little weight, while flexibility, openness to change, and the ability to thrive under varying circumstances are highly prized.

Interdimensional Relations

1. **Cultural Exchanges**: The Realm of Chaos provides unique perspectives on adaptation and survival in extreme conditions, making it a valuable partner in interdimensional research and cultural exchanges. Visitors often come to study its adaptive technologies and philosophical approaches to life.
2. **Impact on Visitors**: Travelers to this realm must be prepared for a world where nothing remains the same for long. The experience can be disorienting but also enlightening, offering a new appreciation for the concepts of order and chaos.

Conclusion

The Realm of Chaos presents a profound exploration of a universe without predictability, where adaptability and flexibility are essential to existence. It challenges visitors and inhabitants alike to reconsider their perceptions of stability and order, providing a unique lens through which to view the potential extremes of universe conditions within the multiverse.

~ 28 ~

THE WORLD OF REFLECTIONS

In this chapter, we venture into the World of Reflections, a dimension where every surface, whether natural or constructed, reflects not only physical appearances but also thoughts, emotions, and past events. This unique environment profoundly impacts the inhabitants' social interactions, cultural developments, and individual identities, offering a comprehensive view of a society deeply intertwined with reflective insight.

Defining Characteristics of the Dimension

1. **Reflective Surfaces as Memory Holders**: In the World of Reflections, mirrors and reflective surfaces are more than simple tools for viewing oneself; they are repositories of history and emotion. These surfaces can display scenes from the past, reflect hidden truths about individuals, or reveal alternate realities, depending on the viewer's state of mind and intent.

2. **Environment and Architecture**: The landscape is dotted with natural reflective pools and crystal structures that store the collective memories of the world's history. Buildings and public spaces are designed with reflective materials that interact dynamically with the populace, changing and

responding to the emotional and psychological states of those around them.

Adaptations to a Reflective World

1. **Psychological Adaptations**: The citizens of the World of Reflections develop high emotional and psychological awareness, necessary to navigate a world where private thoughts can become public knowledge. This transparency fosters a culture of honesty and open communication but also requires careful management of personal emotions and thoughts.
2. **Societal Norms and Privacy**: In a dimension where reflections can reveal one's innermost thoughts and past actions, the concepts of privacy and discretion take on new meanings. Legal and social systems evolve to protect individuals from invasive reflections, balancing the right to privacy with the communal benefits of shared experiences.

Cultural and Social Dynamics

1. **Art and Expression**: Artistic expression in this dimension often involves manipulating reflections to tell stories or convey emotions. Artists might create interactive installations that change based on the viewer's reactions or use reflective surfaces to blend historical and personal narratives.
2. **Education and Learning**: Educational methods leverage reflective technologies to teach history and empathy. Students learn by observing past events and experiencing the emotions of historical figures firsthand, leading to a deep understanding of their culture's heritage and complexity.

Technological Innovations

1. **Reflective Technology**: Technological advancements include the development of reflective devices that can selectively filter or amplify certain memories or emotions. This technology is used in everything from therapy and education to law enforcement and entertainment.

2. **Communication Devices**: Reflective communication devices allow individuals to share experiences and emotions directly, creating a form of empathetic communication that goes beyond words and gestures. This technology enhances social cohesion and understanding but also raises ethical questions about the boundaries of shared experiences.

Philosophical and Ethical Implications

1. **Philosophy of Self and Other**: The omnipresence of reflective surfaces prompts philosophical inquiries into the nature of self, identity, and the other. The constant interplay between one's own reflections and those of others blurs the lines between individual and collective experience.

2. **Ethics of Reflection**: Ethical debates in this world focus on the manipulation of reflections—whether it is moral to alter or suppress certain reflections, and how to balance individual rights with collective needs. These discussions influence policies on the use and regulation of reflective technologies.

Interdimensional Relations

1. **Cultural Exchanges**: The World of Reflections offers unique insights into the nature of memory, perception, and reality, making it a valuable participant in interdimensional studies and cultural exchanges. Visitors from other dimensions often come to explore its reflective arts and to learn its methods of empathetic communication.

2.   **Diplomatic Engagements**: Diplomacy involves complex protocols around the use of reflective surfaces, ensuring that interactions respect the privacy and emotional integrity of all parties involved. These protocols are essential for maintaining trust and openness in interdimensional relations.

Conclusion

The World of Reflections presents a profound exploration of how a society might function when its members cannot hide their true selves and must confront not only their own reflections but those of everyone around them. This dimension challenges us to consider the implications of total transparency and the value of reflection in understanding not only ourselves but also the people with whom we share our worlds.

~ 29 ~

THE GARDEN WORLD

In this chapter, we explore The Garden World, a dimension where the entirety of the planet is covered in lush, verdant gardens and dense, expansive forests. This world is a paradise of biodiversity, where the symbiotic relationship between flora and fauna has led to an ecological utopia. This unique environment provides profound insights into sustainable living and the deep interconnections within an ecosystem.

Environmental Characteristics

1. **Global Flora Coverage**: The Garden World is distinguished by its absence of urban sprawl or barren landscapes. Instead, every inch of the planet is enveloped in various types of gardens, from tropical rainforests and temperate woodlands to extensive flower gardens and vast arboreal expanses.

2. **Climate and Weather**: The climate across this dimension is remarkably stable, moderated by the extensive plant life which regulates atmospheric temperatures and maintains high humidity. Weather patterns are gentle and predictable, supporting a continuous growth cycle without the disruptions of severe weather events.

Ecological Dynamics

1. **Flora and Fauna Interdependence**: The species of The Garden World have evolved in a tightly interwoven web of dependency that enhances the survival of both plant and animal life. Plants evolve complex and often beautiful methods of pollination and seed dispersal, relying on specific animal and insect species whose life cycles are equally tied to these plants.

2. **Advanced Photosynthesis**: Plants in this dimension have developed an advanced form of photosynthesis that allows for more efficient energy storage and greater biomass production. This adaptation supports larger plant and animal species and more vibrant ecosystems than would be possible otherwise.

Societal Structures and Culture

1. **Harmonious Living**: The societies of The Garden World are built around the principles of harmony with nature and sustainability. Cities, if they can be called that, are more like large gardens with homes and buildings seamlessly integrated into the living landscape.

2. **Cultural Significance of Flora**: In this world, plants are not merely background scenery but are revered as vital components of society. Festivals, art, and daily rituals revolve around plant life and the cycles of nature, celebrating the blooming of flowers or the changing colors of leaves.

Technological Innovations

1. **Biological and Botanical Technologies**: Technological advancements in The Garden World are primarily biological and botanical in nature. Bioengineering enhances plant

growth and resilience, developing natural solutions for food production, medicine, and even building materials.

2.   **Transportation and Infrastructure**: Transportation systems are designed to have minimal impact on the environment. Pathways wind through the landscape, and vehicles, often powered by biofuels or solar energy, are designed to blend into the natural surroundings.

Philosophical and Ethical Implications

1.   **Ethics of Coexistence**: Ethical considerations in The Garden World focus on the balance between technological advancement and ecological preservation. The societies here have developed ethical frameworks that prioritize the health of the ecosystem above individual or short-term gains.
2.   **Philosophy of Growth and Renewal**: Philosophical thought in this dimension often reflects themes of growth, renewal, and the cycles of life. There is a deep cultural appreciation for the transient yet perpetual nature of life, mirroring the ever-changing yet enduring garden environment.

Interdimensional Relations

1.   **Cultural Exchanges**: The Garden World is a beacon of ecological and sustainable living, offering valuable lessons in environmental management and harmonious living. Its approach to technology and ethics attracts scholars, environmentalists, and policymakers from various dimensions seeking to learn from its successes.
2.   **Diplomatic Missions**: Diplomatic interactions are conducted with an emphasis on environmental diplomacy, with the Garden World often taking a leading role in interdimensional environmental initiatives and treaties.

Conclusion

The Garden World offers a visionary example of what a society fully integrated with its natural environment might look like, challenging us to reimagine our relationship with nature and the potential for sustainable, symbiotic living. This dimension not only enriches our understanding of ecological balance but also inspires us to incorporate these principles into our own world.

~ 30 ~

THE LABYRINTH DIMENSION

In this chapter, we journey into the Labyrinth Dimension, a world defined by its complex maze-like structures that not only dominate the physical landscape but also shape the cognitive and cultural frameworks of its inhabitants. This dimension is a profound exploration of space, pattern, and the human (and non-human) capacity for problem-solving and orientation.

Geographical and Architectural Features

1. **Endless Maze**: The Labyrinth Dimension consists of an intricate network of interlocking mazes and labyrinths that stretch across the entire planet. These are not merely man-made structures but are formed by natural phenomena such as geological formations, dense forests, and urban sprawls that have evolved into maze-like configurations.

2. **Dynamic Structures**: The labyrinths of this dimension are dynamic and ever-changing. Walls and paths may shift, driven by natural processes or by the labyrinths' own inexplicable logic, making navigation challenging and requiring constant adaptation.

Adaptations to Maze Life

1.   **Navigational Skills**: The inhabitants of the Labyrinth Dimension have developed exceptional navigational skills. Early education focuses heavily on orientation, memory, and spatial awareness. Residents often use a combination of traditional compasses, celestial navigation, and unique bio-sensory abilities to find their way.

2.   **Flora and Fauna**: The wildlife has also adapted to the labyrinthine environment. Many creatures are equipped with heightened senses such as echolocation or sensitive vibrational perception to navigate and communicate through the complex corridors.

Societal Structures and Culture

1.   **Community and Isolation**: Societies within the Labyrinth Dimension are often isolated from each other by vast and confusing mazes, leading to a rich diversity of independent cultures. However, large gatherings are held at the centers of labyrinths during certain periods when paths are known to open, fostering a sense of community and shared heritage.

2.   **Cultural Significance of the Maze**: In this dimension, the maze is a powerful symbol of life's journey and individual growth. It represents the challenges each person must navigate, both physically and spiritually, and is a common motif in art, literature, and religion.

Technological Innovations

1.   **Navigation Technologies**: Technology in this dimension includes sophisticated navigation aids, ranging from advanced GPS-like systems adapted to the shifting labyrinths to personal drones that scout paths and map changes in real time.

2. **Architectural Techniques**: Building technologies are highly advanced, with structures designed to be as flexible and adaptable as the labyrinths themselves. Buildings can expand, contract, or shift as needed, incorporating smart materials that respond to environmental conditions.

Philosophical and Ethical Implications

1. **Philosophy of Pathfinding**: Philosophically, the Labyrinth Dimension revolves around concepts of destiny, free will, and the paths one chooses in life. The labyrinth serves as a metaphor for the choices and challenges faced by all beings, emphasizing the belief in each individual's journey through existence.

2. **Ethics of Space and Movement**: Ethical issues often arise concerning the rights to alter or navigate spaces, with significant legal and moral considerations governing who can modify or set the routes within the labyrinths. These rights impact everything from daily travel to larger questions about societal control and freedom.

Interdimensional Relations

1. **Cultural Exchanges**: Despite its inherent challenges, the Labyrinth Dimension is part of a vibrant network of interdimensional trade and cultural exchange, offering unique insights into architecture, philosophy, and technology.

2. **Diplomatic Challenges**: Diplomatic missions to and from this dimension require careful planning and local guidance to navigate the labyrinthine barriers, highlighting the need for cooperation and mutual understanding in overcoming physical and metaphorical obstacles.

Conclusion

The Labyrinth Dimension offers a unique lens through which to view the complexities of life and society. It challenges its inhabitants—and us, as observers—to consider the ways in which environments shape cultures and cognitive abilities. The labyrinth, as both reality and metaphor, provides a rich ground for exploring themes of journey, challenge, and triumph in the face of adversity.

~ 31 ~

THE OCEAN SKY

In this chapter, we traverse the mesmerizing domain of the Ocean Sky, a dimension where the sky and sea are inverted, creating a world where oceans float above and landscapes stretch below. This inversion of sea and sky challenges conventional perceptions and physics, fostering unique ecosystems and advanced aerial and aquatic adaptations.

Defining the Inversion of Sky and Sea

1. **Aerial Oceans**: In the Ocean Sky dimension, vast bodies of water hover above the earth's surface, suspended by unique gravitational anomalies that reverse the typical behavior of water. These aerial oceans teem with marine life and are dotted with floating islands that serve as habitats for various species.

2. **Terrestrial Landscapes Below**: Beneath the floating oceans, the landscapes resemble expansive valleys and plains that receive moisture from the "rain" falling from the oceans above. These regions support diverse terrestrial ecosystems and are crucial for agriculture and habitation.

Adaptations to an Inverted Environment

1. **Aerial Marine Life**: Marine creatures in this dimension have evolved to navigate and thrive in the suspended oceans. Many species have developed lighter-than-water bodies or air-filled bladders that allow them to float and move through the aerial seas effortlessly.

2. **Skyfaring Flora**: Flora has adapted to both float in the aerial oceans and root on the floating islands. Some plants have developed buoyant seeds that travel via air currents between islands, while others rely on water birds and flying fish for pollination and seed dispersal.

Societal Structures and Cultural Dynamics

1. **Skyborne Cities**: Human and sentient inhabitants live on floating islands and construct their cities to withstand the movements of the aerial oceans. Buildings are designed with buoyant materials and flexible foundations, allowing them to adapt to the shifting tides above.

2. **Cultural Significance of Water and Air**: In this dimension, water and air are not seen as separate elements but as interconnected forces of life. Cultural and spiritual practices often celebrate the fluidity and unity of these elements, with festivals and rituals that emphasize their cyclical and life-sustaining nature.

Technological Innovations

1. **Navigation and Transport**: Transportation technologies are highly advanced, with vehicles designed for seamless travel between the aerial oceans and the land below. These include hybrid air-water crafts capable of navigating both air currents and water flows.

2. **Resource Management**: Ingenious methods are developed for harvesting water and nutrients from the aerial oceans.

Technologies such as airborne aqueducts and automated drones are used to collect water and fish, which are vital resources for the population below.

Philosophical and Ethical Implications

1. **Philosophy of Balance and Unity**: The philosophy in the Ocean Sky dimension centers around concepts of balance, unity, and adaptability. The inhabitants view their lives as perpetually connected between the sky and the earth, fostering a deep respect for their environment and its unique challenges.

2. **Ethics of Environmental Interaction**: Ethical considerations focus on maintaining the delicate balance between aerial and terrestrial ecosystems. Laws and guidelines strictly regulate environmental modifications, ensuring that the symbiotic relationship between the sky oceans and the land is preserved.

Interdimensional Relations

1. **Cultural Exchanges**: The Ocean Sky dimension is a hub for scholars and environmentalists interested in its unique ecological model and sustainable practices. Its approach to living in harmony with a radically different environment provides valuable lessons for other worlds.

2. **Diplomatic Engagements**: Diplomatic missions from the Ocean Sky dimension often involve discussions on environmental sustainability and technological innovations, offering insights into living harmoniously within challenging environments.

Conclusion

The Ocean Sky dimension redefines the boundaries of natural law and the possibilities of life. It challenges us to rethink our relationship with the natural world and inspires innovative approaches to sustainability and coexistence. This dimension not only expands our understanding of ecological interdependence but also illuminates the profound connections between water and air as essential elements of life.

~ 32 ~

CONCLUSION: INTEGRATING THE MULTIVERSE INTO OUR LIVES

As we conclude our exploration of the multiverse and its myriad dimensions, we reflect on the profound lessons and insights gleaned from the diverse realities we've encountered. Each dimension, from the Ocean Sky to the Labyrinth Dimension, from realms governed by musical harmony to those dictated by the laws of chaos, offers a unique perspective on existence, challenging our preconceptions and expanding our understanding of what it means to be part of this vast universe.

Reflections on Parallel Dimensions

The journey through the multiverse reveals not just the scientific possibilities of other worlds but also the philosophical and spiritual insights they provide. We've seen dimensions where time flows backward, where landscapes are alive with sound, and where societies are built on principles vastly different from our own. These explorations encourage us to question the very nature of reality, pushing the boundaries of our imagination and scientific understanding.

Each dimension, with its unique laws and societal structures, serves as a mirror reflecting our own world's possibilities and limitations. By examining the ways in which these alternate universes operate, we gain a deeper appreciation for the complexity and diversity of the cosmos. It reminds us that our reality is just one of many, each with its own lessons and mysteries.

Enhancing Personal Growth and Understanding

The knowledge of parallel dimensions can profoundly impact personal growth and worldview. Recognizing the vastness of possibilities helps cultivate an open mind, a critical asset in a world that is increasingly interconnected and complex. It encourages flexibility in thought and a readiness to accept and adapt to change—qualities essential for personal and collective evolution.

Furthermore, understanding that there are multiple ways of existing and solving problems inspires creativity and innovation. It challenges us to think outside the conventional frameworks and consider alternative solutions to global challenges, be they environmental, social, or technological.

Encouraging Openness to Infinite Possibilities

The final and perhaps most crucial takeaway from our journey through the multiverse is the encouragement to remain open to infinite possibilities. Just as the dimensions we explored are diverse and bound by different rules, so too can our perspectives be flexible and inclusive. This openness is vital not only in scientific and philosophical inquiries but also in everyday interactions and decisions.

As we integrate the lessons from the multiverse into our lives, we are reminded of the importance of diversity, the strength of adaptability, and the power of seeing beyond our immediate reality. We learn that each moment, each decision, echoes across not just our world but potentially across an infinite array of universes, impacting the cosmos in ways we can only begin to imagine.

Final Thoughts

Let us carry forward the curiosity and awe that the concept of the multiverse inspires. Let these explorations remind us of our place within a larger cosmos and motivate us to live with intention, creativity, and an enduring sense of wonder. May the knowledge of the multiverse inspire us to dream bigger, act with compassion and curiosity, and embrace the mysteries that lie ahead.

As we close this book, remember that the journey through the multiverse does not end here. It extends into our lives, our dreams, and our actions. The multiverse invites us to continually explore, question, and expand the horizons of our understanding. Let us step forward with open minds, ready to embrace the infinite possibilities that await.

~ 33 ~

APPENDIX: FURTHER READING
AND RESOURCES

This appendix provides a curated list of books, articles, documentaries, and practices designed to deepen your understanding of quantum physics, metaphysics, the multiverse, and ways to explore connections with other dimensions through meditation and psychic exercises.

Books on Quantum Physics and the Multiverse

1. **"The Elegant Universe" by Brian Greene** - A comprehensive and engaging exploration of superstring theory and its implications for the concept of the multiverse.
2. **"Fabric of the Cosmos" by Brian Greene** - Explores the texture and structure of the universe, explaining concepts of space, time, and the very nature of reality.
3. **"The Hidden Reality: Parallel Universes and the Deep Laws of the Cosmos" by Brian Greene** - A detailed discussion on the various multiverse theories and the deep laws that govern our universe.
4. **"Quantum Enigma: Physics Encounters Consciousness" by Bruce Rosenblum and Fred Kuttner** - Explores the

fascinating implications of quantum theory and its intersection with consciousness studies.

5.   **"Warped Passages: Unraveling the Mysteries of the Universe's Hidden Dimensions" by Lisa Randall** - Discusses the possibility and implications of other dimensions existing alongside ours.

Documentaries and Video Series

1.   **"The Fabric of the Cosmos" by NOVA** - A visually stunning series that brings to life the scientific concepts and mysteries of the universe.

2.   **"Particle Fever"** - Follows six scientists during the launch of the Large Hadron Collider, providing a dramatic and exciting look at the big bang and the conditions of the early universe.

3.   **"What the Bleep Do We Know!?"** - This film explores the connections between quantum mechanics, neurobiology, human consciousness, and reality through a blend of documentary interviews and narrative storytelling.

Articles and Online Resources

1.   **"Exploring the Multiverse" by Sean Carroll** - A compelling article that discusses the scientific and philosophical questions surrounding the multiverse.

2.   **ArXiv.org** - An open-access archive for scholarly articles in physics, mathematics, and computer science, where many papers on quantum physics and the multiverse are available for free.

3.   **"Multiverse Theories: A Philosophical Perspective" by Stanford Encyclopedia of Philosophy** - Offers a philosophical analysis of the multiverse theories, providing a deeper understanding of their implications and the ongoing debates.

Meditation and Psychic Exercises

1. **Guided Meditation for Multiverse Exploration** - Look for guided meditations that focus on expanding consciousness and exploring parallel dimensions. These often involve visualization techniques and are available through meditation apps or YouTube.

2. **Psychic Development Exercises:**

   - *Astral Projection Practice* - Begin with guided sessions that help you learn to safely experience and navigate astral projection, a state where the consciousness travels outside the physical body.
   - *Intuitive Reading Sessions* - Practice tuning into intuitive information about yourself and others, which can be a step toward sensing information across dimensions.

3. **Workshops and Retreats:**

   - Look for workshops that focus on quantum healing, energy work, and psychic development. These often provide practical tools and community support for exploring connections with other dimensions.

4. **Daily Journaling:**

   - Keep a dream journal to track your dreams and explore possible connections or messages from other dimensions.
   - Regularly journal about your meditative experiences to deepen your understanding and recall of the experiences.

Conclusion

This list of resources is meant to guide you through the complex theories of quantum physics and the metaphysical aspects of the multiverse while also providing practical tools for personal exploration. Whether through reading, viewing, or direct practice, these resources can expand your understanding and perhaps even alter your perception of reality.

~ 34 ~

GLOSSARY OF TERMS

This glossary provides definitions of key scientific and metaphysical terms used throughout the book to enhance understanding and clarity of complex concepts related to the multiverse, quantum physics, and metaphysical explorations.

1. Astral Projection:

- *Definition*: A form of out-of-body experience where the astral body (spirit or consciousness) separates from the physical body to travel in the astral plane, which is thought to be an intermediate world of various dimensions.

2. Bioluminescence:

- *Definition*: The production and emission of light by a living organism, typically through a chemical reaction involving luciferin and luciferase. Bioluminescence is used by several species for attraction, camouflage, and communication.

3. Crystal Dimension:

- *Definition*: A hypothetical parallel universe where crystalline structures dominate the ecosystem and technology, influencing the physical and social environment significantly.

4. Dimension:

- *Definition*: In the context of this book, a dimension is an alternate realm or universe with properties and laws that may differ significantly from those of our known universe.

5. Echolocation:

- *Definition*: A biological sonar used by several kinds of animals (such as bats and dolphins) to navigate and locate objects by emitting sounds and listening for the echoes that return from them.

6. Entropy:

- *Definition*: A measure of disorder or randomness in a system. In thermodynamics, it is a central concept used to explain the direction of spontaneous processes.

7. Multiverse:

- *Definition*: A theoretical framework in physics and cosmology that posits the existence of many universes, including the observable universe in which we live. These universes may vary in terms of laws of physics and natural constants.

8. Quantum Mechanics:

- *Definition*: A fundamental theory in physics that provides a description of the physical properties of nature at the scale of atoms and subatomic particles.

9. Quantum Superposition:

- *Definition*: A fundamental principle of quantum mechanics that posits a physical system (such as an electron) exists partly in all its particular, theoretically possible states simultaneously; however, when measured, it gives a result corresponding to one of the possible configurations.

10. T-Symmetry (Time Reversal Symmetry):

- *Definition*: A symmetry of the fundamental physical laws under the reversal of time direction. When applied, the laws that describe the motion of particles in motion should also apply if the direction of time is reversed.

11. Telepathy:

- *Definition*: The purported transmission of information from one person to another without using any known human sensory channels or physical interaction.

12. Theoretical Physics:

- *Definition*: A branch of physics that employs mathematical models and abstractions to explain and predict natural phenomena, focusing on the development and logical implications of quantum theory, relativity, and other high-level theories.

13. Warp Drive:

- *Definition*: A speculative idea based on the solution of Einstein's field equations in general relativity, suggesting a method of faster-than-light travel by expanding spacetime behind a spaceship and contracting spacetime in front.

14. Wormhole:

- *Definition*: A hypothetical topological feature of spacetime that would fundamentally be a 'shortcut' through spacetime. A wormhole is envisioned as a tunnel with two ends at separate points in spacetime.

15. Zen Meditation:

- *Definition*: A type of meditation that originated in Buddhism and focuses on gaining insight into the nature of existence and personal awakening, often used to enhance mindfulness and emotional calm.

This glossary aims to assist readers in navigating the complex concepts discussed in the book and to provide a clearer understanding of the scientific and metaphysical discussions that underpin the exploration of parallel dimensions.

EPILOGUE

As we conclude our journey through the vast and variegated landscapes of the multiverse, I, Demetri Welsh, wish to thank you for joining me on this expedition beyond the familiar confines of our known universe. If our travels through these pages have seemed at times as much a philosophical inquiry as a scientific exploration, it is because the notion of parallel realities challenges not only our understanding of the cosmos but also our conception of ourselves.

Throughout *Parallel Lives, Parallel Realities: Unveiling the Multiverse*, we have traversed dimensions where the laws of physics diverge wildly from our own, where the fundamental elements of existence—time, space, matter, and energy—play out in novel and unexpected ways. We've imagined worlds that defy gravity, embrace the eternal, and live in perpetual shadow or unending light. Each chapter, each dimension, was crafted not just to entertain but to provoke thought, to expand our understanding of what might be possible in the vast theater of the cosmos.

As your guide, I have aimed to blend the rigor of scientific speculation with the boundless possibilities of metaphysical exploration. By marrying these two realms, we have not only envisioned alternate universes but have also reflected on the profound questions they raise about fate, free will, consciousness, and the very nature of reality itself. These explorations are more than just mental exercises; they are a reminder of the inherent mystery that underpins our existence. They prompt us to consider that perhaps, in some hidden fold of the cosmos, versions of us are looking up at the same stars, pondering similar profound questions.

If this book has succeeded in its purpose, you will close its covers with a sense that reality is richer and more layered than it appears; that beyond every sunset, there may lie infinite sunrises in countless other worlds; that every choice we make echoes across possible universes, crafting shadow stories that we might one day learn to read.

But beyond the intellectual enrichment such thoughts provide, there is a deeper, more personal transformation I hope you take with you—an openness to the unknown, a readiness to question what seems unquestionable, and a renewed wonder at the universe and our place within it.

About The Author

Demetri Welsh is a celebrated author and online psychic known for his profound insight and transformative guidance in the realm of spiritual discovery. He has dedicated his career to helping individuals explore their psychic potential and deepen their connection with the metaphysical world. His unique approach combines accessibility with deep personal engagement, providing insights that are not only about prediction but are also tools for self-discovery and personal growth.

Demetri's written works, including titles like *Unlocking Your Inner Vision: A Teen's Guide to Psychic Development* and *Mystic Whispers: A Beginner's Journey into Spellcraft and Psychic Exploration*, cater to both young adults and seasoned seekers alike. These books are crafted to guide readers through various psychic practices ranging from tarot reading to energy work, aiming to awaken their innate intuitive abilities and spiritual awareness.

In his practice, Demetri offers personalized sessions that extend beyond geographical boundaries, thanks to the digital nature of his services. He is recognized for his ability to create intimate and comfortable virtual spaces for his clients, facilitating emotional healing, spiritual enlightenment, and practical advice for navigating life's complexities.

Through his empathetic and ethically grounded approach, Demetri Welsh continues to be a guiding light for those on a journey towards greater understanding and enlightenment in the ever-evolving spiritual landscape.

www.DemetriWelsh.com